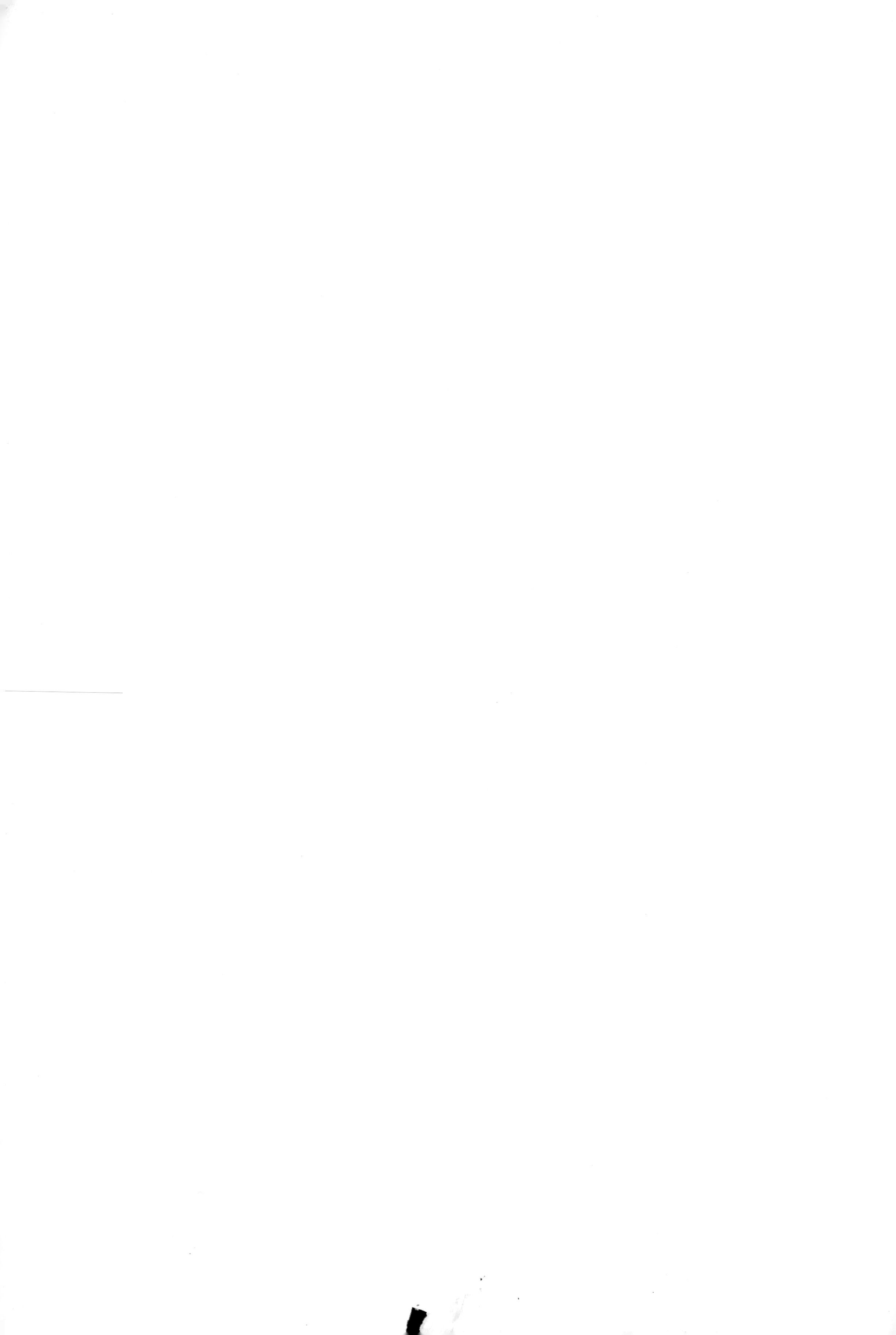

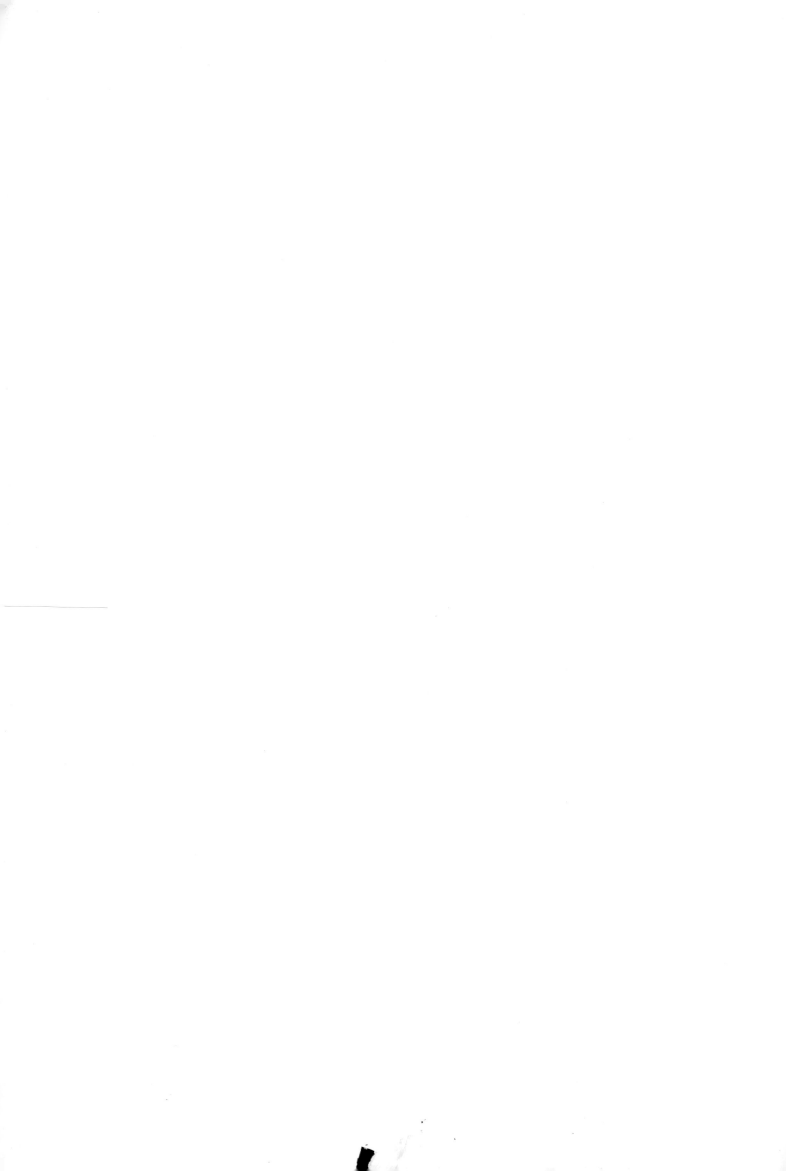

ATLAS OF ANATOMY

Original title of the book in Spanish: *Atlas de Anatomía*.
© Copyright PARRAMÓN EDICIONES, S. A. 1995—World Rights.
Published by Parramón Ediciones, S. A., Barcelona, Spain.

Author: Parramón's Editorial Team.
Illustrator: Antonio Muñoz Tenllado.
© Copyright of the English edition: BARRON'S EDUCATIONAL SERIES, INC. 1997
Translator: Ana María Soler-Rodríguez, Ph.D.

All inquiries should be addressed to:
Barron's Educational Series, Inc.
250 Wireless Boulevard
Hauppauge, New York 11788

International Standard Book Number 0-7641-5000-6
Library of Congress Catalog Card Number 96-79470

Printed in Spain

9 8 7 6 5 4 3

ATLAS OF
ANATOMY

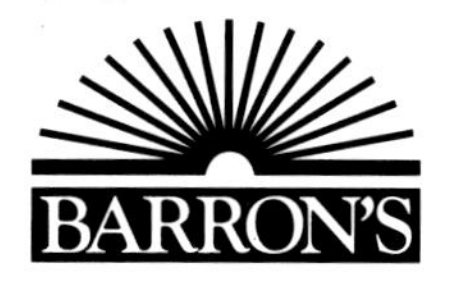

BARRON'S

Synopsis

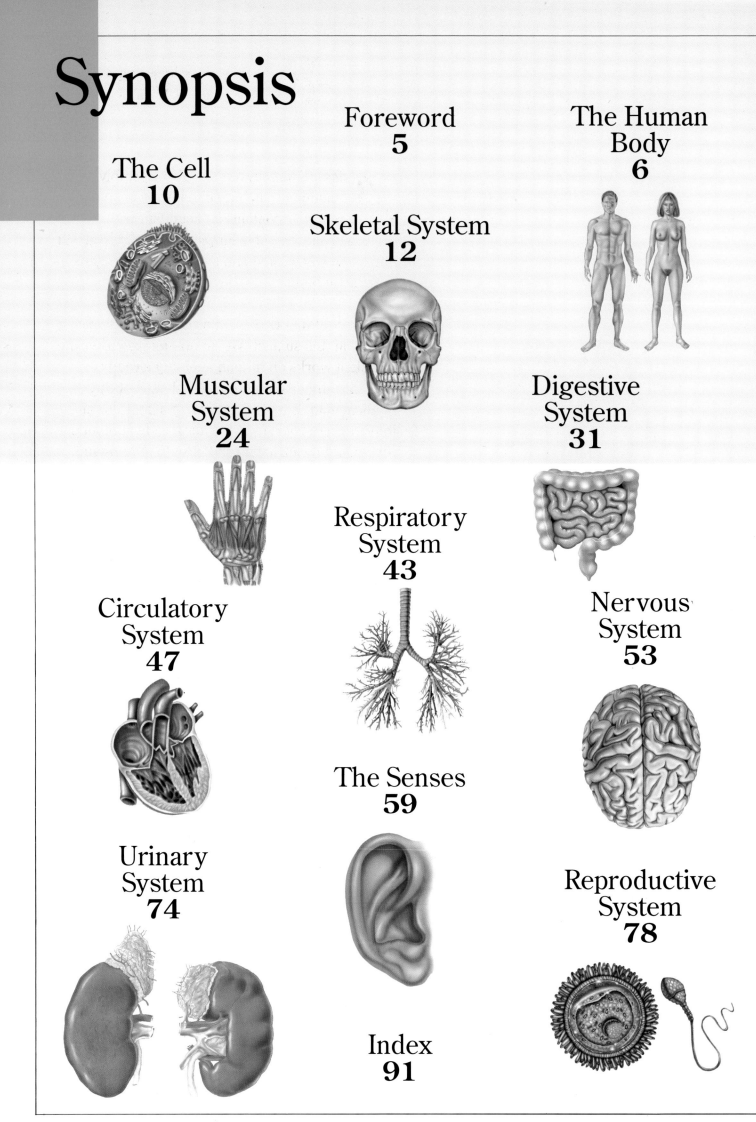

Foreword
5

The Human
Body
6

The Cell
10

Skeletal System
12

Muscular
System
24

Digestive
System
31

Respiratory
System
43

Circulatory
System
47

Nervous
System
53

The Senses
59

Urinary
System
74

Reproductive
System
78

Index
91

Foreword

*A*tlas of Anatomy outlines the assembly and function of the human body. As in any atlas, this work contains an abundant and selective collection of color plates in which you will find, correctly located and labeled, all the organs of the human body.

The present market is well provided with basic books on this subject. We do not ignore either the excellent works on human anatomy already available and destined to professionals of the medical field. But we are missing a book for an intermediate level, destined for a wide audience, that would be attractive and easy to understand.

This has been precisely the target area for *Atlas of Anatomy*. By providing more information than the basic books but less than those books dealing with specialized medicine, we plan to satisfy the needs of a large number of readers.

Atlas of Anatomy will be a good working and study tool for high school students, as well as for people interested in medicine or professionally related to it (students of medicine or physical education, medical assistants, and so forth). It should also be useful to people interested in acquiring a general knowledge of anatomy.

The objective of this book is to help every person who wants to know more about this wonderful machine called the human body.

The Human Body
1. Male Surface Anatomy

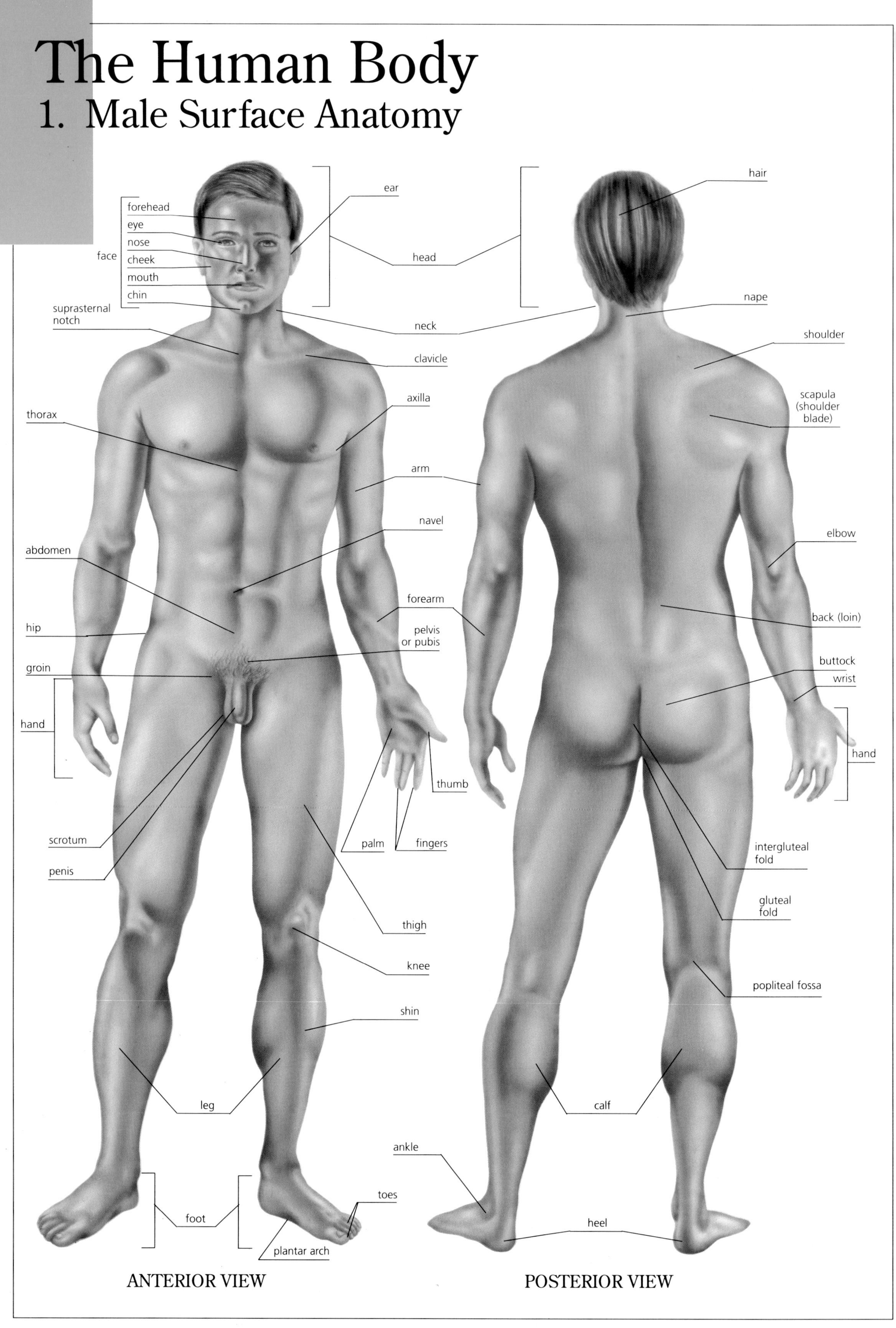

ear

forehead

eye

nose

face

cheek

mouth

chin

head

hair

suprasternal notch

neck

nape

clavicle

shoulder

thorax

axilla

scapula (shoulder blade)

arm

navel

elbow

abdomen

hip

back (loin)

groin

pelvis or pubis

forearm

buttock

wrist

hand

scrotum

penis

thumb

palm

fingers

intergluteal fold

gluteal fold

thigh

knee

popliteal fossa

shin

calf

leg

ankle

toes

foot

heel

plantar arch

ANTERIOR VIEW

POSTERIOR VIEW

4. Regions of the Body. Posterior View

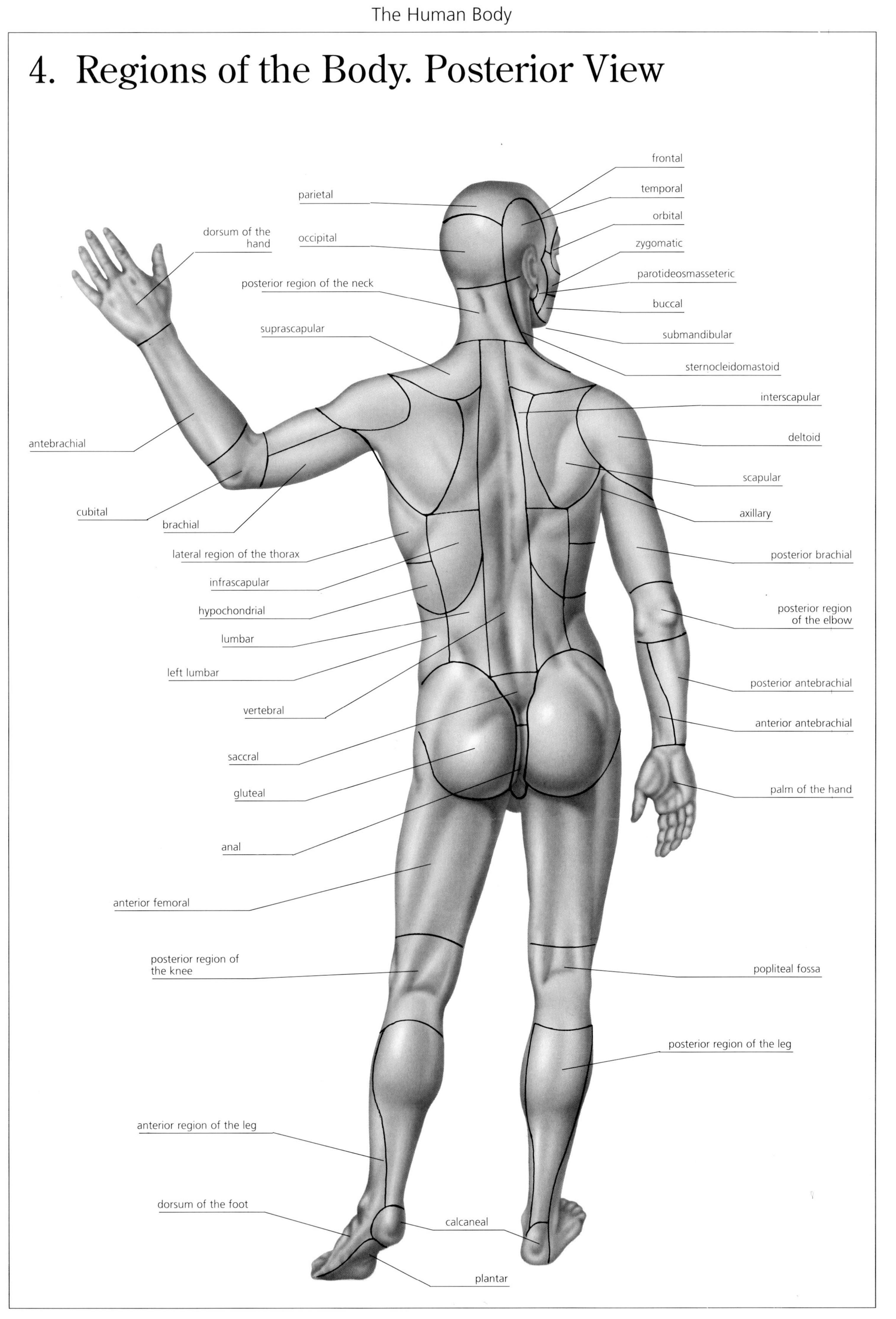

parietal

dorsum of the
hand

occipital

posterior region of the neck

suprascapular

antebrachial

cubital

brachial

lateral region of the thorax

infrascapular

hypochondrial

lumbar

left lumbar

vertebral

saccral

gluteal

anal

anterior femoral

posterior region of
the knee

anterior region of the leg

dorsum of the foot

frontal

temporal

orbital

zygomatic

parotideosmasseteric

buccal

submandibular

sternocleidomastoid

interscapular

deltoid

scapular

axillary

posterior brachial

posterior region
of the elbow

posterior antebrachial

anterior antebrachial

palm of the hand

popliteal fossa

posterior region of the leg

calcaneal

plantar

The Cell
1. The Human Cell

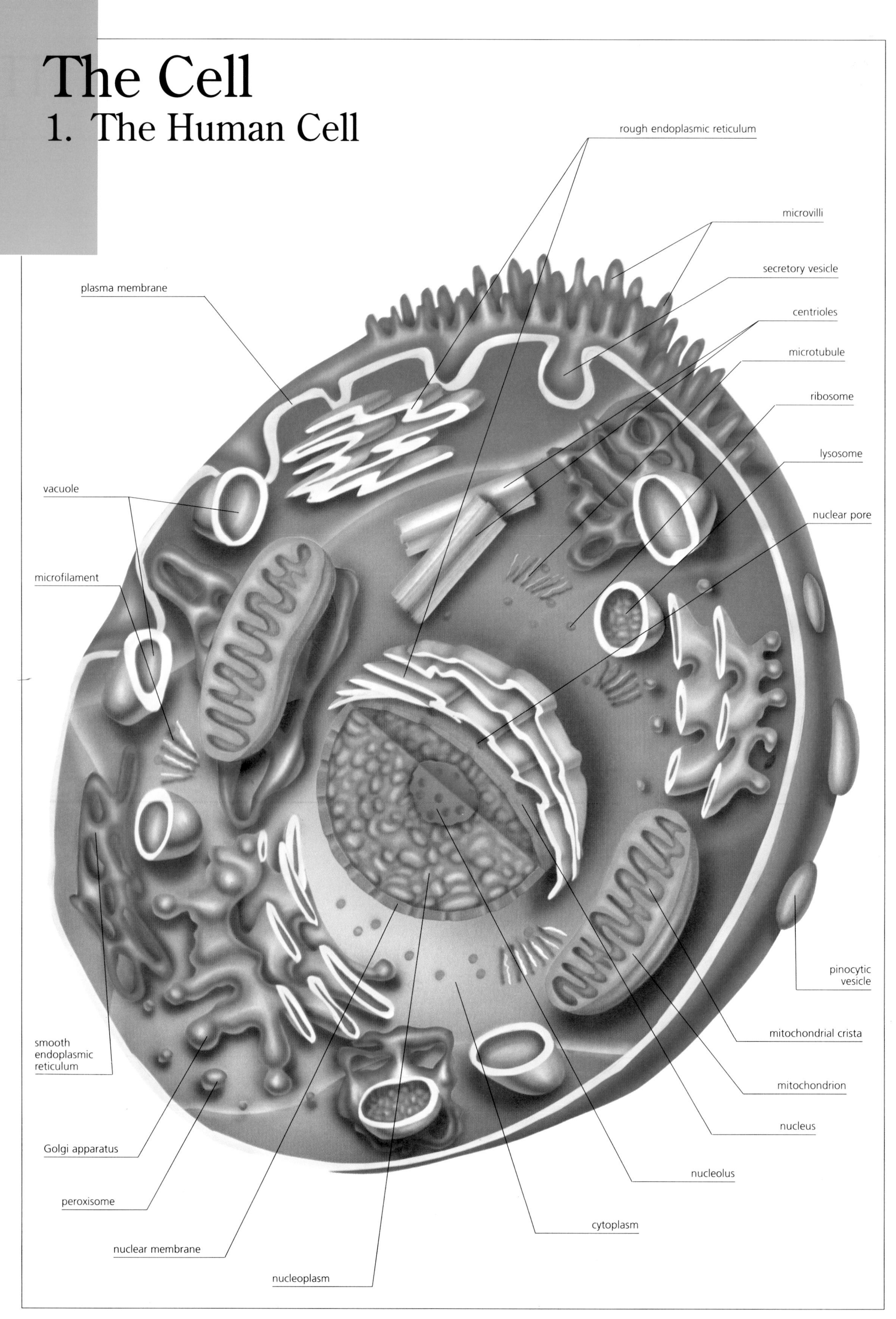

rough endoplasmic reticulum

microvilli

secretory vesicle

centrioles

microtubule

ribosome

lysosome

nuclear pore

plasma membrane

vacuole

microfilament

pinocytic vesicle

mitochondrial crista

mitochondrion

nucleus

smooth endoplasmic reticulum

Golgi apparatus

peroxisome

nucleolus

cytoplasm

nuclear membrane

nucleoplasm

2. Types of Cells. DNA

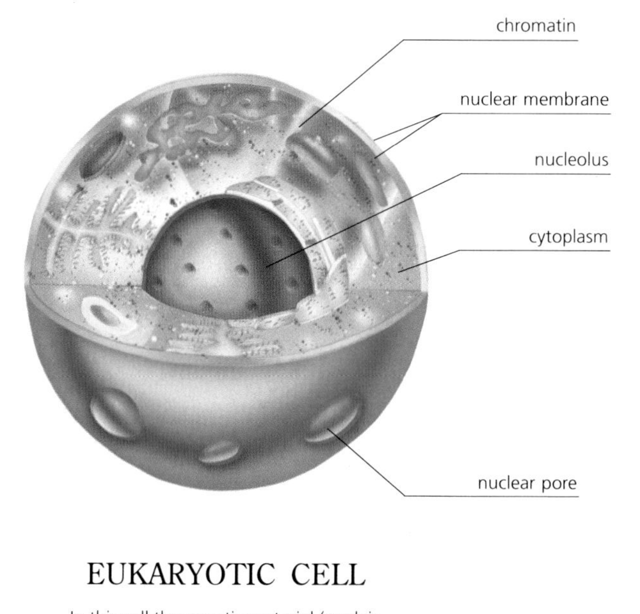

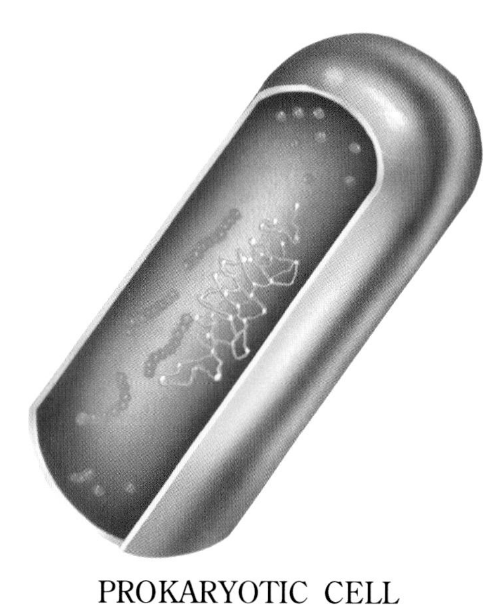

chromatin

nuclear membrane

nucleolus

cytoplasm

nuclear pore

EUKARYOTIC CELL

In this cell the genetic material (nucleic acid) is in the nucleus. Typical of human cells.

PROKARYOTIC CELL

In this cell the genetic material (nucleic acid) is free in the cytoplasm. Bacteria and blue-green algae.

GLYCOGEN

(long chain of glucose molecules)

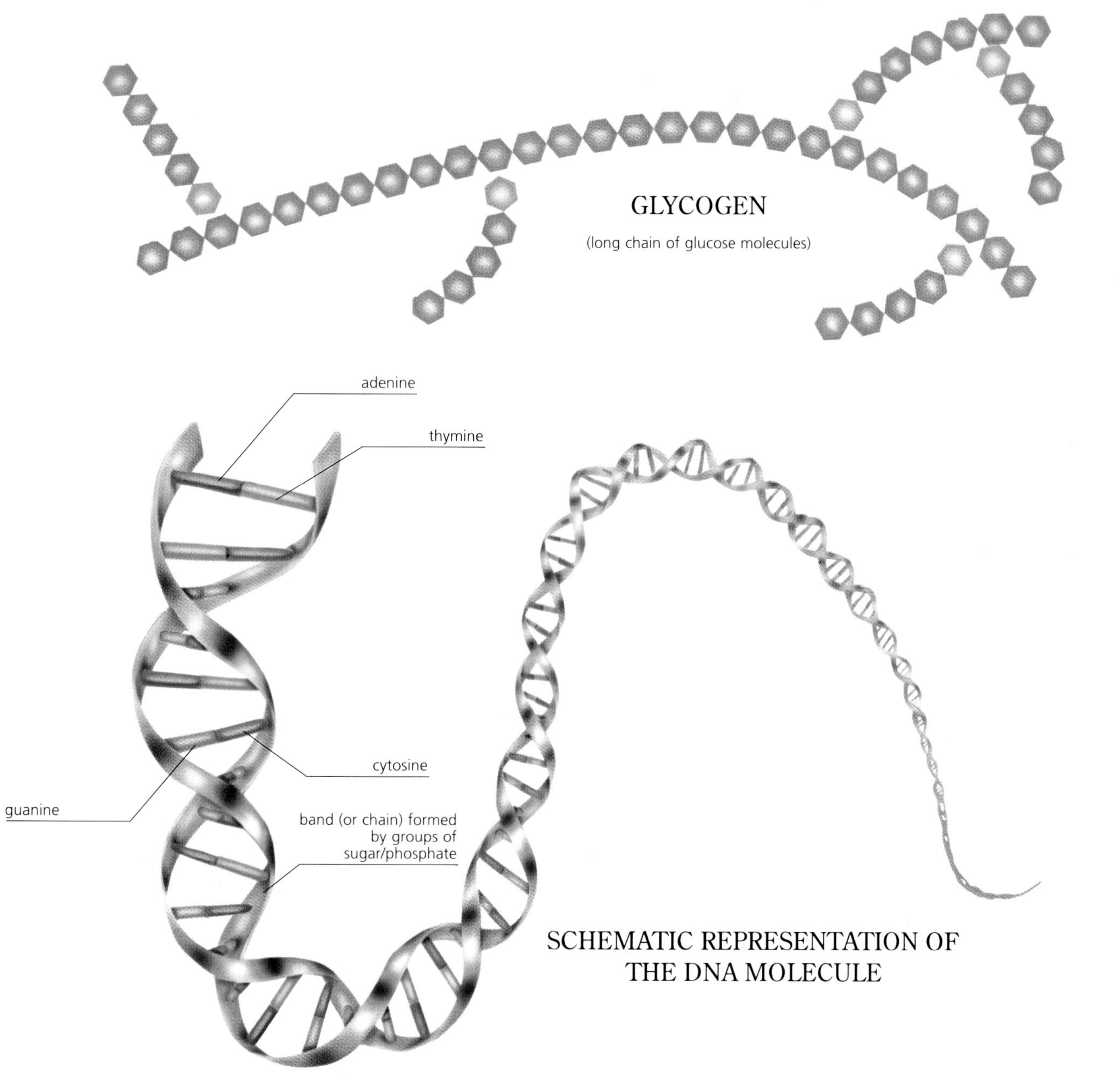

adenine

thymine

cytosine

guanine

band (or chain) formed by groups of sugar/phosphate

SCHEMATIC REPRESENTATION OF THE DNA MOLECULE

The Skeletal System
1. Bones of the Head

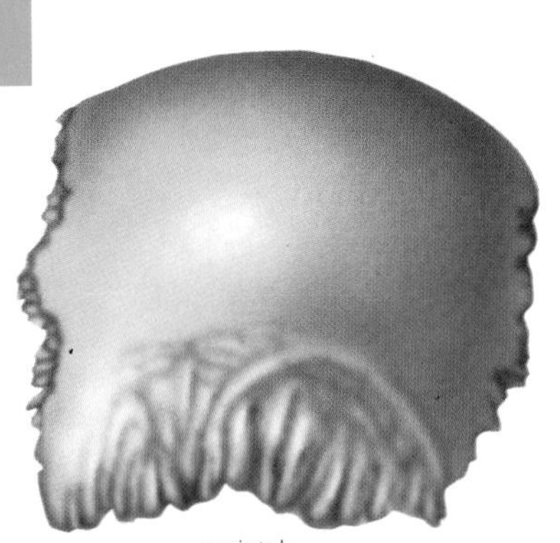

parietal

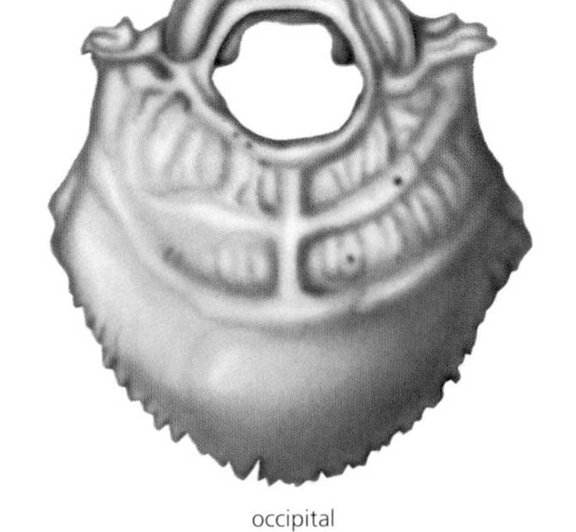

occipital

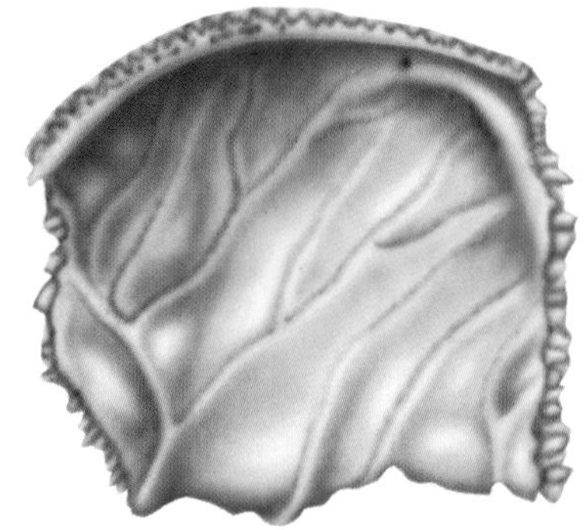

parietal

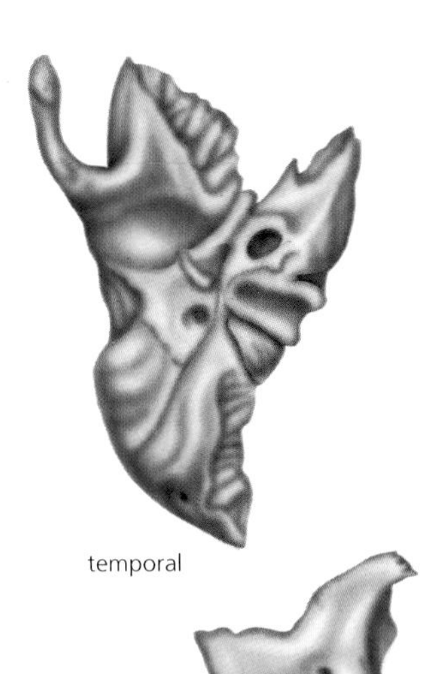

temporal

frontal

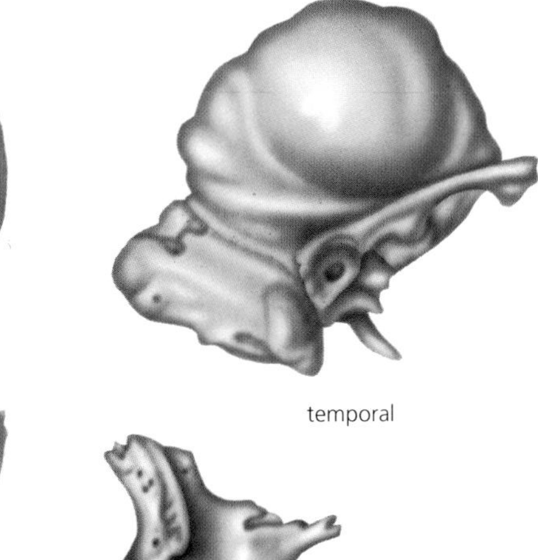

temporal

zygomatic

sphenoid

zygomatic

lacrimal

lacrimal

incus

malleus

stapes

palatine

ethmoid

palatine

vomer

stapes

malleus

incus

inferior concha

nasal

inferior concha

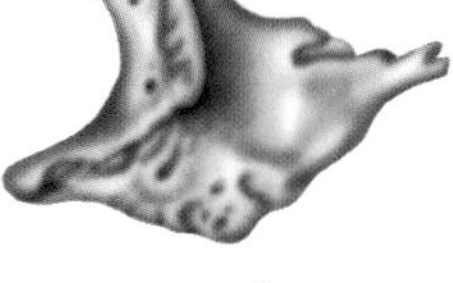

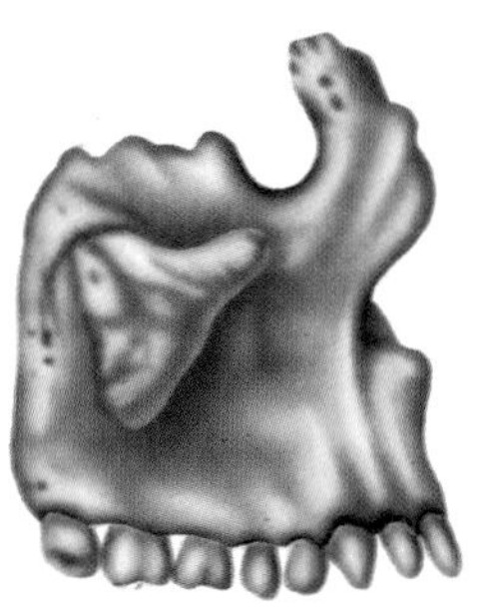

maxilla

hyoid

maxilla

mandible

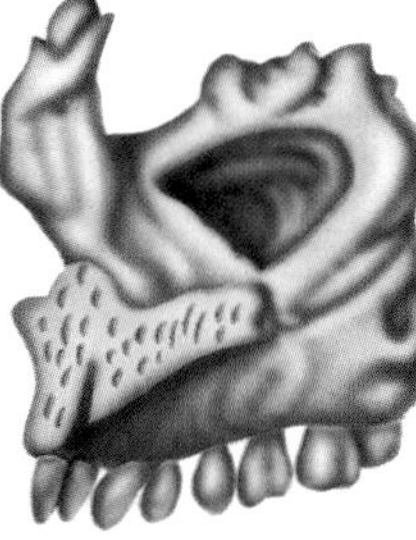

2. Bones of the Trunk and Limbs

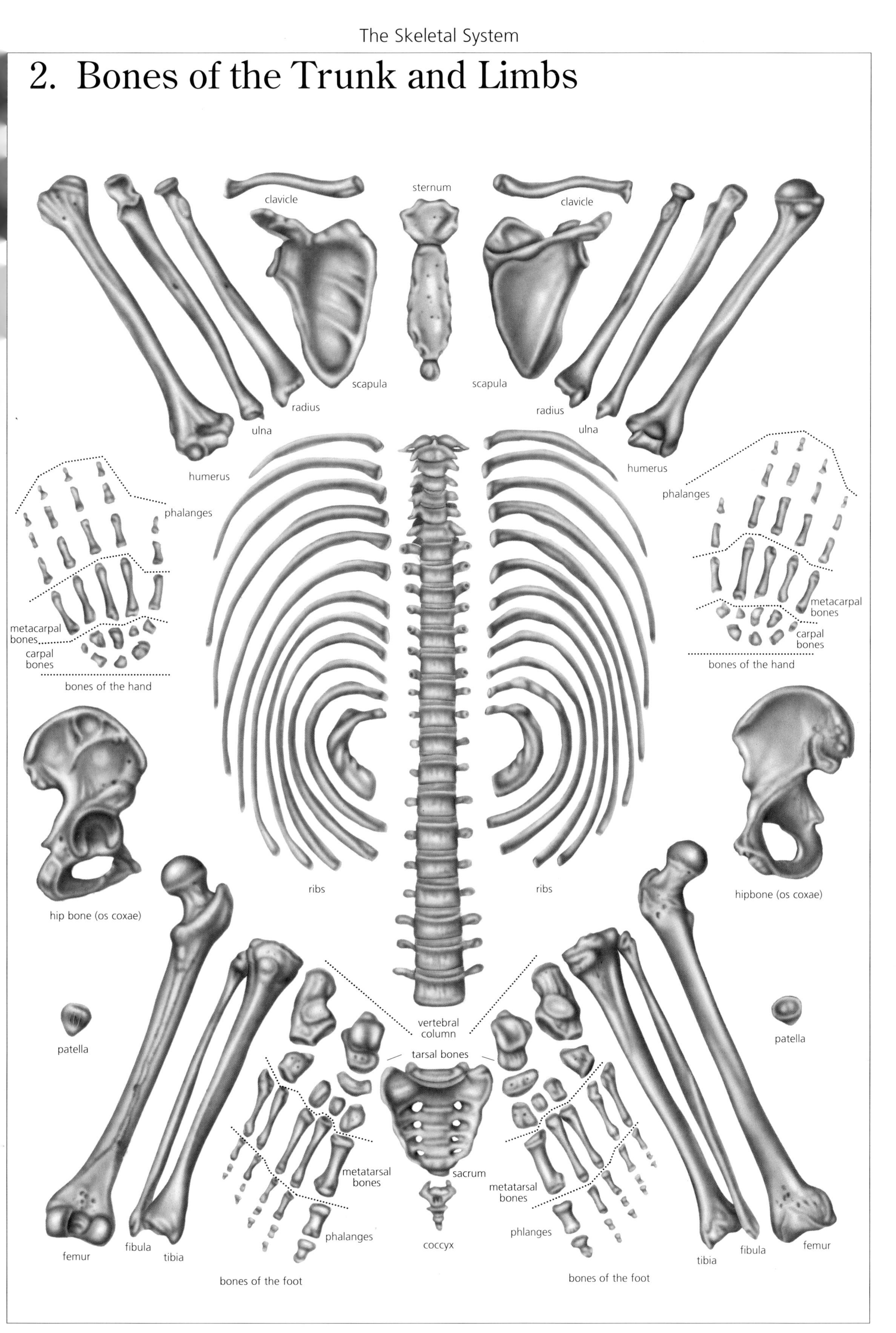

clavicle

sternum

clavicle

scapula

scapula

radius

radius

ulna

ulna

humerus

humerus

phalanges

phalanges

phalanges

metacarpal bones

metacarpal bones

carpal bones

carpal bones

bones of the hand

bones of the hand

ribs

ribs

hip bone (os coxae)

hipbone (os coxae)

vertebral column

tarsal bones

patella

patella

metatarsal bones

metatarsal bones

sacrum

femur

fibula

tibia

phalanges

coccyx

phlanges

tibia

fibula

femur

bones of the foot

bones of the foot

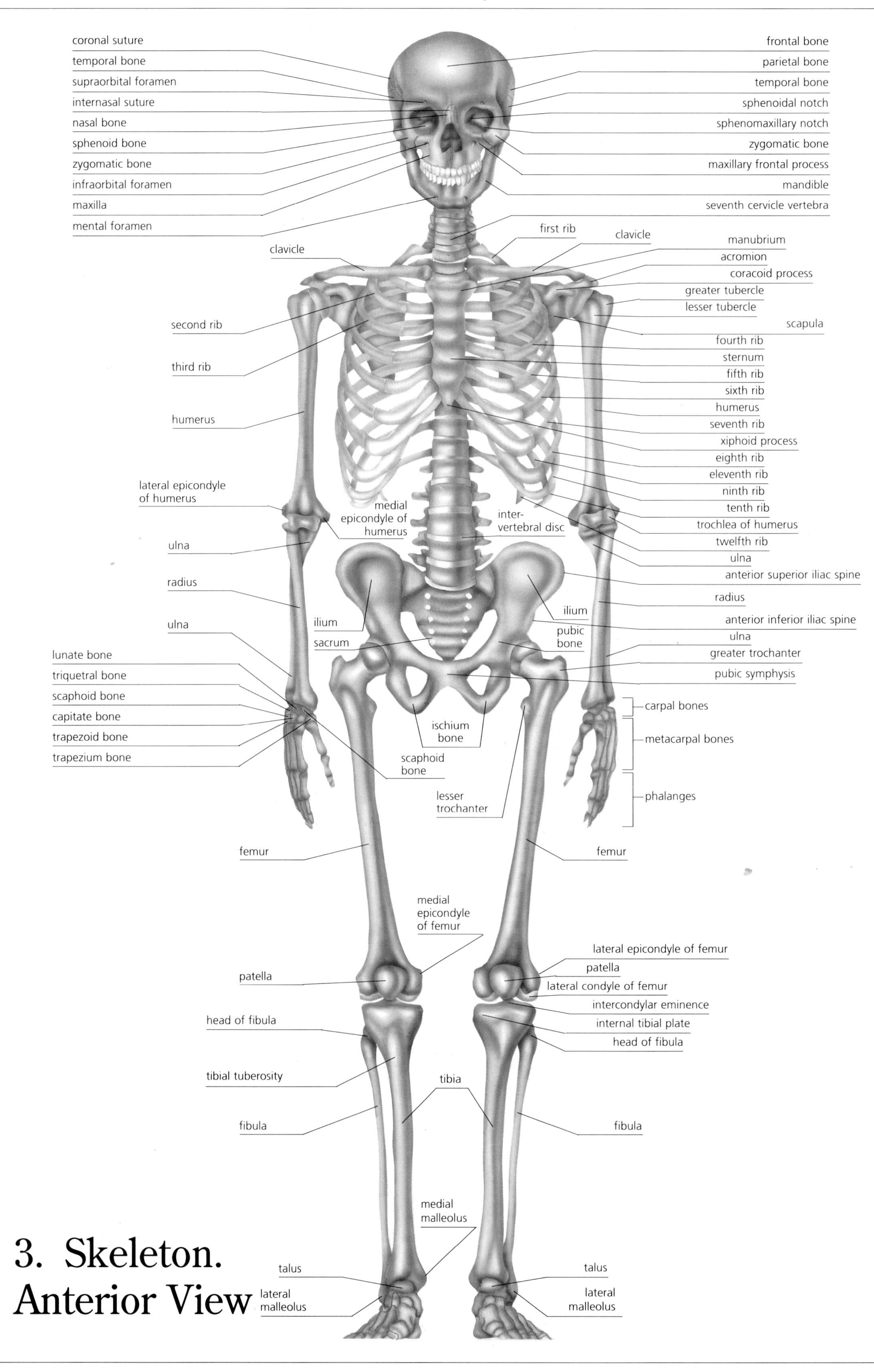

coronal suture
temporal bone
supraorbital foramen
internasal suture
nasal bone
sphenoid bone
zygomatic bone
infraorbital foramen
maxilla
mental foramen

frontal bone
parietal bone
temporal bone
sphenoidal notch
sphenomaxillary notch
zygomatic bone
maxillary frontal process
mandible
seventh cervicle vertebra

clavicle
second rib
third rib
humerus

first rib   clavicle
                   manubrium
                   acromion
                   coracoid process
                   greater tubercle
                   lesser tubercle
                        scapula
                   fourth rib
                   sternum
                   fifth rib
                   sixth rib
                   humerus
                   seventh rib
                   xiphoid process
                   eighth rib
                   eleventh rib
                   ninth rib
                   tenth rib
                   trochlea of humerus
                   twelfth rib
                   ulna
                   anterior superior iliac spine
                   radius
                   anterior inferior iliac spine
                   ulna
                   greater trochanter
                   pubic symphysis

lateral epicondyle
of humerus

medial
epicondyle of
humerus

inter-
vertebral disc

ulna

radius

ulna

ilium
sacrum

ilium
pubic
bone

lunate bone
triquetral bone
scaphoid bone
capitate bone
trapezoid bone
trapezium bone

carpal bones

metacarpal bones

phalanges

ischium
bone

scaphoid
bone

lesser
trochanter

femur

femur

medial
epicondyle
of femur

patella

head of fibula

tibial tuberosity

fibula

lateral epicondyle of femur
patella
lateral condyle of femur
intercondylar eminence
internal tibial plate
head of fibula

tibia

fibula

medial
malleolus

talus
lateral
malleolus

talus
lateral
malleolus

3. Skeleton.
Anterior View

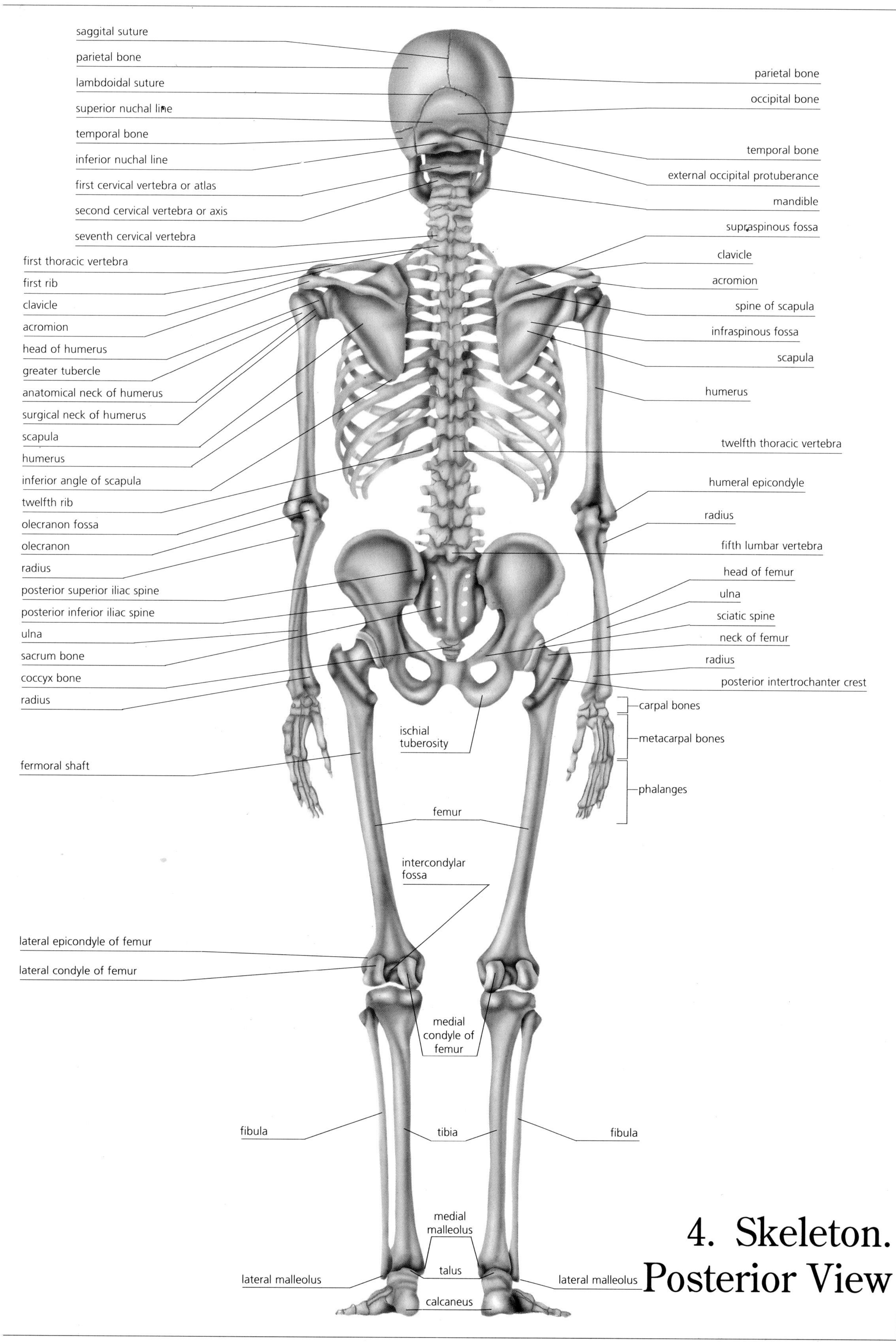

saggital suture

parietal bone

lambdoidal suture

superior nuchal line

temporal bone

inferior nuchal line

first cervical vertebra or atlas

second cervical vertebra or axis

seventh cervical vertebra

first thoracic vertebra

first rib

clavicle

acromion

head of humerus

greater tubercle

anatomical neck of humerus

surgical neck of humerus

scapula

humerus

inferior angle of scapula

twelfth rib

olecranon fossa

olecranon

radius

posterior superior iliac spine

posterior inferior iliac spine

ulna

sacrum bone

coccyx bone

radius

fermoral shaft

lateral epicondyle of femur

lateral condyle of femur

parietal bone

occipital bone

temporal bone

external occipital protuberance

mandible

supraspinous fossa

clavicle

acromion

spine of scapula

infraspinous fossa

scapula

humerus

twelfth thoracic vertebra

humeral epicondyle

radius

fifth lumbar vertebra

head of femur

ulna

sciatic spine

neck of femur

radius

posterior intertrochanter crest

carpal bones

metacarpal bones

phalanges

ischial tuberosity

femur

intercondylar fossa

medial condyle of femur

fibula

tibia

fibula

medial malleolus

lateral malleolus

talus

lateral malleolus

calcaneus

4. Skeleton.
Posterior View

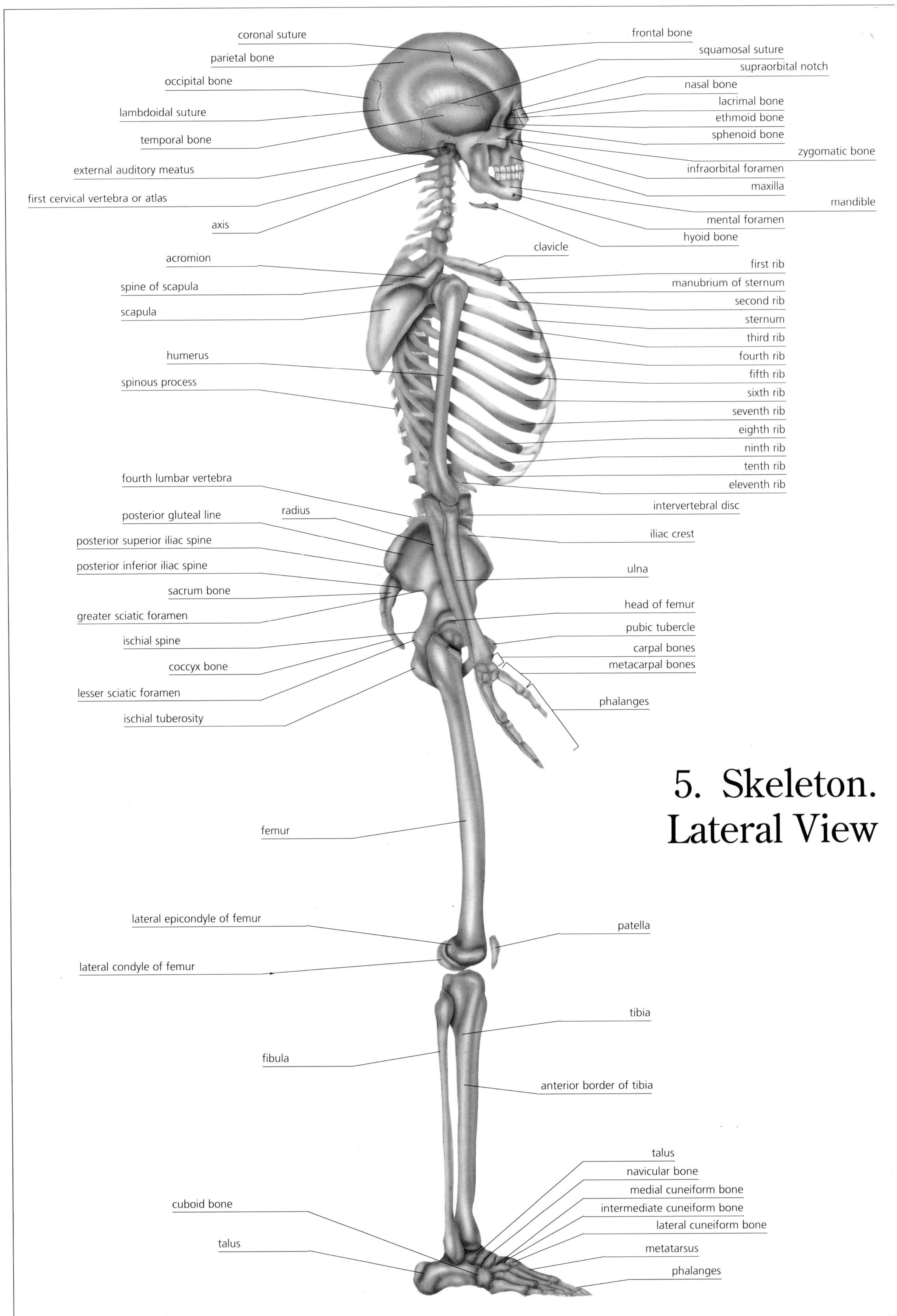

coronal suture
parietal bone
occipital bone
lambdoidal suture
temporal bone
external auditory meatus
first cervical vertebra or atlas
axis
acromion
spine of scapula
scapula
humerus
spinous process
fourth lumbar vertebra
posterior gluteal line
posterior superior iliac spine
posterior inferior iliac spine
sacrum bone
greater sciatic foramen
ischial spine
coccyx bone
lesser sciatic foramen
ischial tuberosity
radius
femur
lateral epicondyle of femur
lateral condyle of femur
fibula
cuboid bone
talus

frontal bone
squamosal suture
supraorbital notch
nasal bone
lacrimal bone
ethmoid bone
sphenoid bone
zygomatic bone
infraorbital foramen
maxilla
mandible
mental foramen
hyoid bone
clavicle
first rib
manubrium of sternum
second rib
sternum
third rib
fourth rib
fifth rib
sixth rib
seventh rib
eighth rib
ninth rib
tenth rib
eleventh rib
intervertebral disc
iliac crest
ulna
head of femur
pubic tubercle
carpal bones
metacarpal bones
phalanges
patella
tibia
anterior border of tibia
talus
navicular bone
medial cuneiform bone
intermediate cuneiform bone
lateral cuneiform bone
metatarsus
phalanges

5. Skeleton. Lateral View

6. Skull

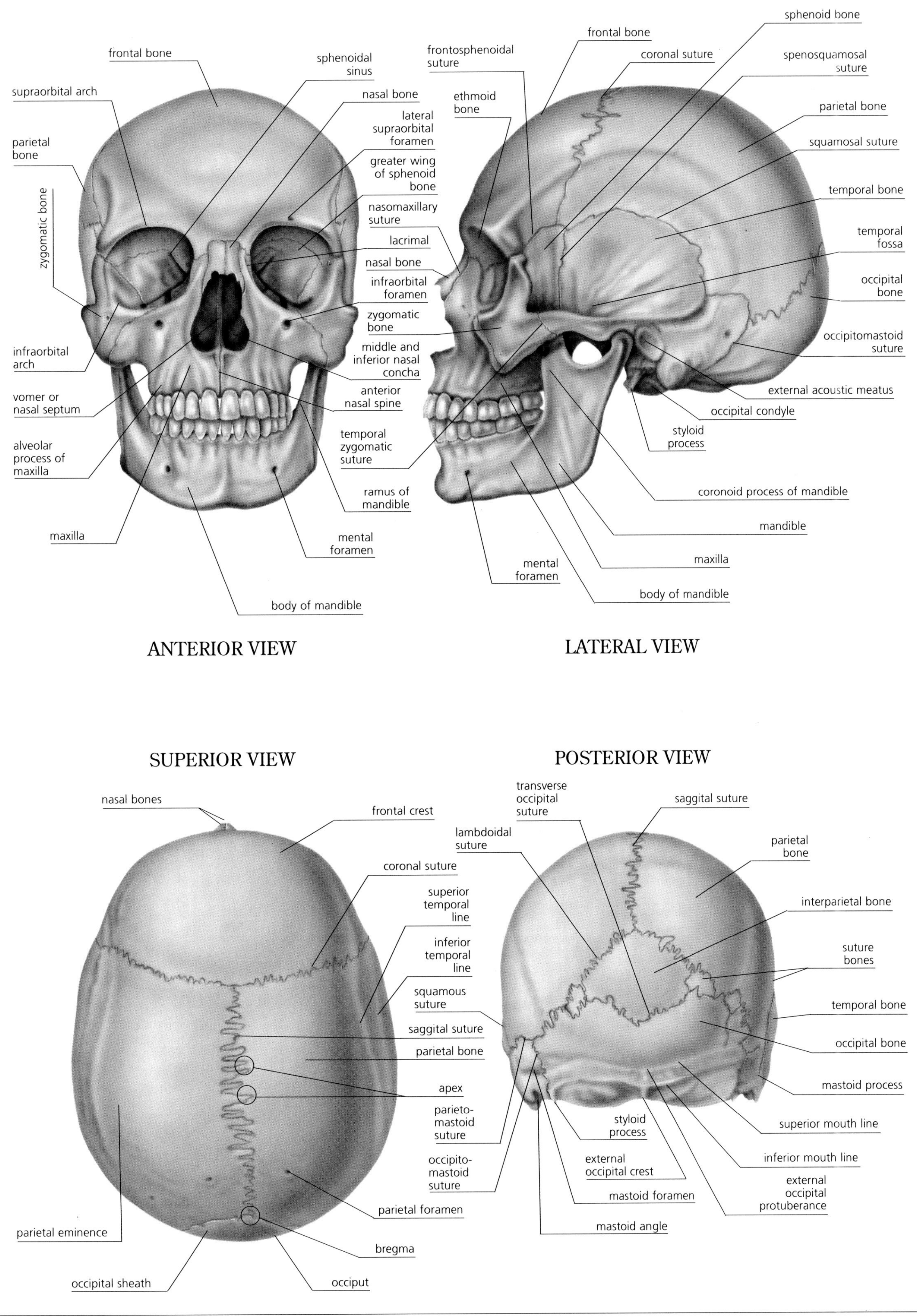

ANTERIOR VIEW

frontal bone

supraorbital arch

parietal bone

zygomatic bone

infraorbital arch

vomer or nasal septum

alveolar process of maxilla

maxilla

sphenoidal sinus

nasal bone

lateral supraorbital foramen

greater wing of sphenoid bone

nasomaxillary suture

lacrimal

nasal bone

infraorbital foramen

zygomatic bone

middle and inferior nasal concha

anterior nasal spine

temporal zygomatic suture

ramus of mandible

mental foramen

body of mandible

LATERAL VIEW

frontosphenoidal suture

ethmoid bone

frontal bone

coronal suture

sphenoid bone

spenosquamosal suture

parietal bone

squamosal suture

temporal bone

temporal fossa

occipital bone

occipitomastoid suture

external acoustic meatus

occipital condyle

styloid process

coronoid process of mandible

mandible

maxilla

body of mandible

mental foramen

SUPERIOR VIEW

nasal bones

frontal crest

coronal suture

superior temporal line

inferior temporal line

squamous suture

saggital suture

parietal bone

apex

parieto-mastoid suture

occipito-mastoid suture

parietal foramen

bregma

occiput

occipital sheath

parietal eminence

POSTERIOR VIEW

transverse occipital suture

lambdoidal suture

saggital suture

parietal bone

interparietal bone

suture bones

temporal bone

occipital bone

mastoid process

superior mouth line

inferior mouth line

external occipital protuberance

mastoid angle

mastoid foramen

external occipital crest

styloid process

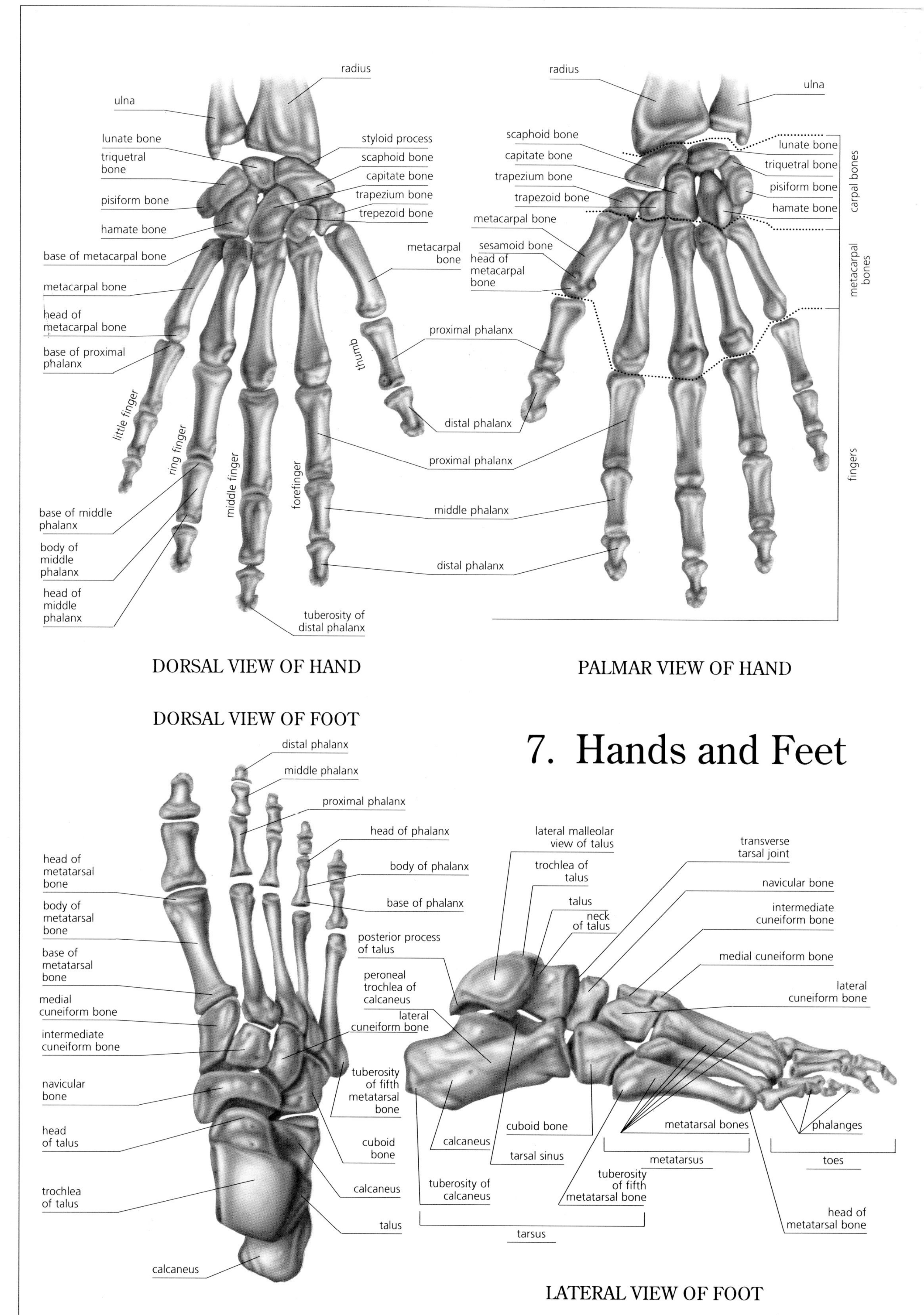

radius

ulna

lunate bone

triquetral bone

pisiform bone

hamate bone

base of metacarpal bone

metacarpal bone

head of metacarpal bone

base of proximal phalanx

styloid process

scaphoid bone

capitate bone

trapezium bone

trepezoid bone

metacarpal bone

little finger

ring finger

middle finger

forefinger

thumb

base of middle phalanx

body of middle phalanx

head of middle phalanx

tuberosity of distal phalanx

DORSAL VIEW OF HAND

radius

ulna

scaphoid bone

capitate bone

trapezium bone

trapezoid bone

metacarpal bone

sesamoid bone

head of metacarpal bone

lunate bone

triquetral bone

pisiform bone

hamate bone

carpal bones

metacarpal bones

proximal phalanx

distal phalanx

proximal phalanx

middle phalanx

distal phalanx

fingers

PALMAR VIEW OF HAND

DORSAL VIEW OF FOOT

7. Hands and Feet

distal phalanx

middle phalanx

proximal phalanx

head of phalanx

body of phalanx

base of phalanx

head of metatarsal bone

body of metatarsal bone

base of metatarsal bone

medial cuneiform bone

intermediate cuneiform bone

navicular bone

head of talus

trochlea of talus

posterior process of talus

peroneal trochlea of calcaneus

lateral cuneiform bone

tuberosity of fifth metatarsal bone

cuboid bone

calcaneus

talus

calcaneus

lateral malleolar view of talus

trochlea of talus

talus

neck of talus

calcaneus

tarsal sinus

tuberosity of calcaneus

cuboid bone

tuberosity of fifth metatarsal bone

transverse tarsal joint

navicular bone

intermediate cuneiform bone

medial cuneiform bone

lateral cuneiform bone

metatarsal bones

phalanges

metatarsus

toes

head of metatarsal bone

tarsus

LATERAL VIEW OF FOOT

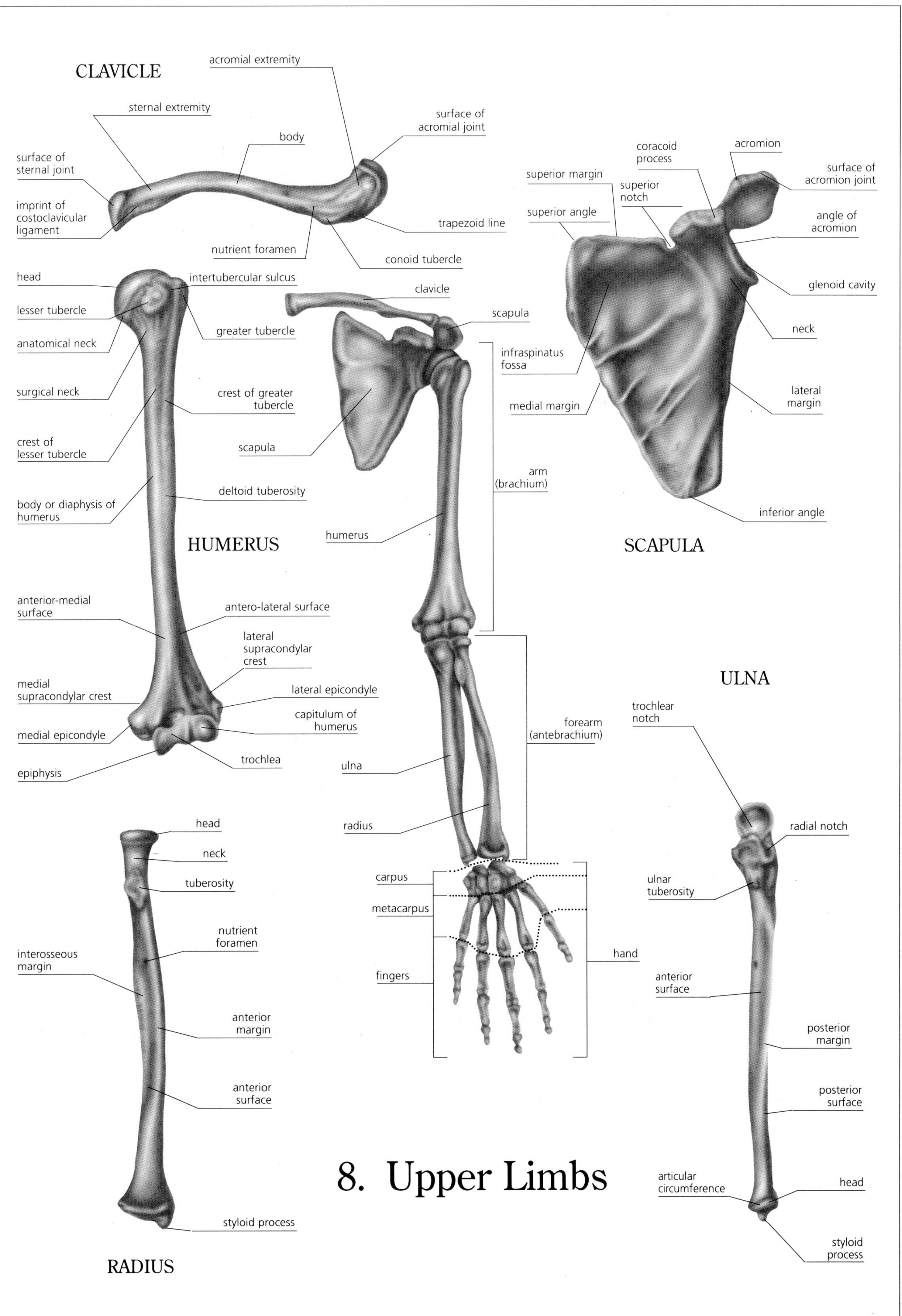

CLAVICLE

acromial extremity

sternal extremity

body

surface of
acromial joint

surface of
sternal joint

imprint of
costoclavicular
ligament

nutrient foramen

trapezoid line

conoid tubercle

head

intertubercular sulcus

lesser tubercle

greater tubercle

anatomical neck

surgical neck

crest of greater
tubercle

crest of
lesser tubercle

scapula

body or diaphysis of
humerus

deltoid tuberosity

HUMERUS

clavicle

scapula

infraspinatus
fossa

medial margin

arm
(brachium)

humerus

coracoid
process

acromion

superior margin

superior
notch

surface of
acromion joint

superior angle

angle of
acromion

glenoid cavity

neck

lateral
margin

inferior angle

SCAPULA

anterior-medial
surface

antero-lateral surface

lateral
supracondylar
crest

medial
supracondylar crest

lateral epicondyle

capitulum of
humerus

medial epicondyle

epiphysis

trochlea

ulna

radius

forearm
(antebrachium)

ULNA

trochlear
notch

radial notch

ulnar
tuberosity

head

neck

tuberosity

carpus

metacarpus

nutrient
foramen

fingers

hand

anterior
surface

interosseous
margin

posterior
margin

anterior
margin

posterior
surface

anterior
surface

articular
circumference

head

styloid process

styloid
process

RADIUS

8. Upper Limbs

9. Lower Limbs

HIP BONE (OS COXAE)

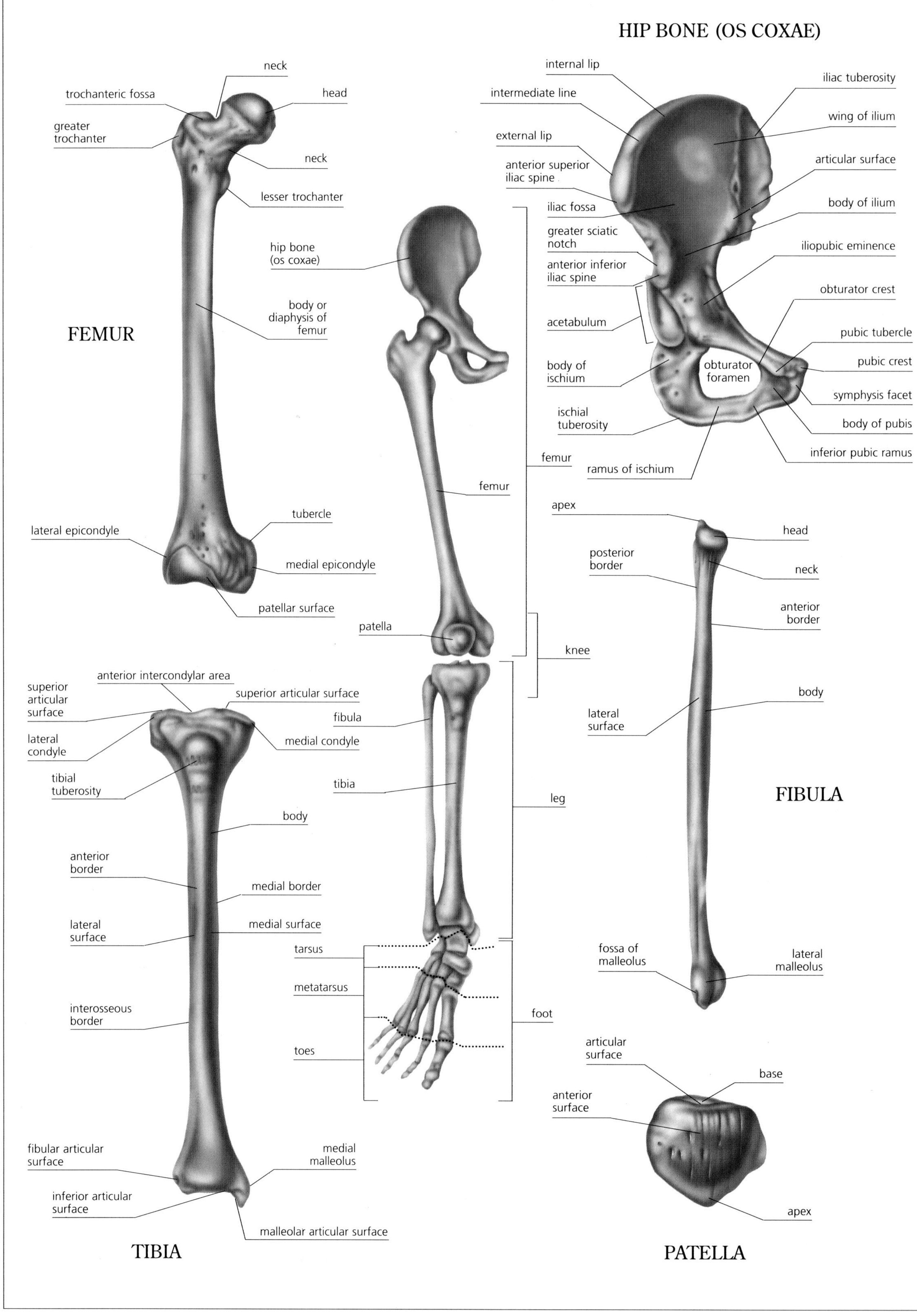

neck

trochanteric fossa

head

greater trochanter

neck

lesser trochanter

hip bone (os coxae)

body or diaphysis of femur

FEMUR

lateral epicondyle

tubercle

medial epicondyle

patellar surface

internal lip

intermediate line

external lip

anterior superior iliac spine

iliac fossa

greater sciatic notch

anterior inferior iliac spine

acetabulum

body of ischium

ischial tuberosity

femur

iliac tuberosity

wing of ilium

articular surface

body of ilium

iliopubic eminence

obturator crest

pubic tubercle

obturator foramen

pubic crest

symphysis facet

body of pubis

inferior pubic ramus

ramus of ischium

femur

apex

head

posterior border

neck

anterior border

patella

knee

body

lateral surface

anterior intercondylar area

superior articular surface

superior articular surface

fibula

lateral condyle

medial condyle

tibial tuberosity

tibia

body

anterior border

medial border

medial surface

lateral surface

leg

FIBULA

tarsus

metatarsus

foot

interosseous border

fossa of malleolus

lateral malleolus

toes

articular surface

base

fibular articular surface

medial malleolus

anterior surface

inferior articular surface

malleolar articular surface

apex

TIBIA

PATELLA

10. Vertebral Column

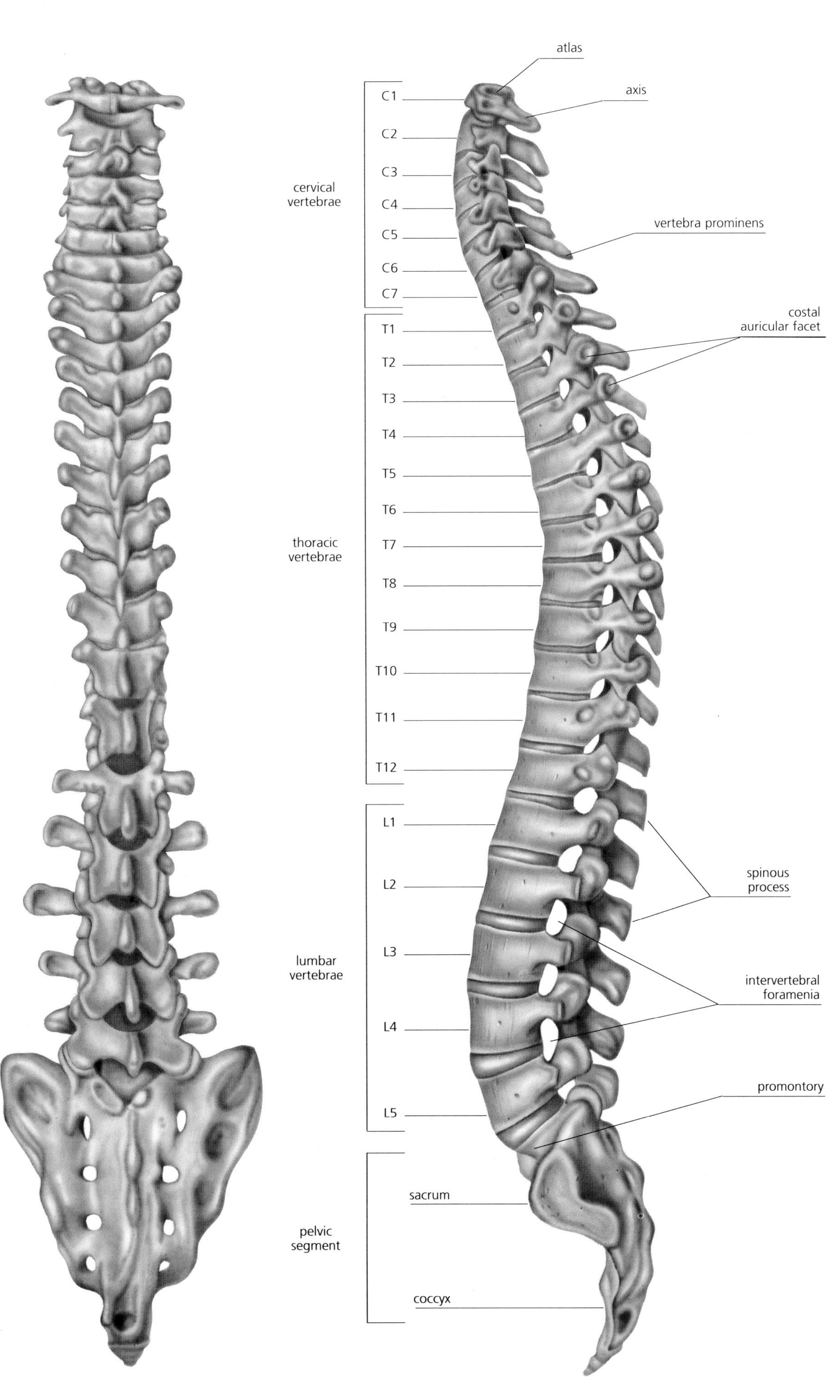

cervical vertebrae

C1
C2
C3
C4
C5
C6
C7

atlas
axis

vertebra prominens

thoracic vertebrae

T1
T2
T3
T4
T5
T6
T7
T8
T9
T10
T11
T12

costal auricular facet

lumbar vertebrae

L1
L2
L3
L4
L5

spinous process

intervertebral foramenia

promontory

pelvic segment

sacrum

coccyx

11. Vertebrae

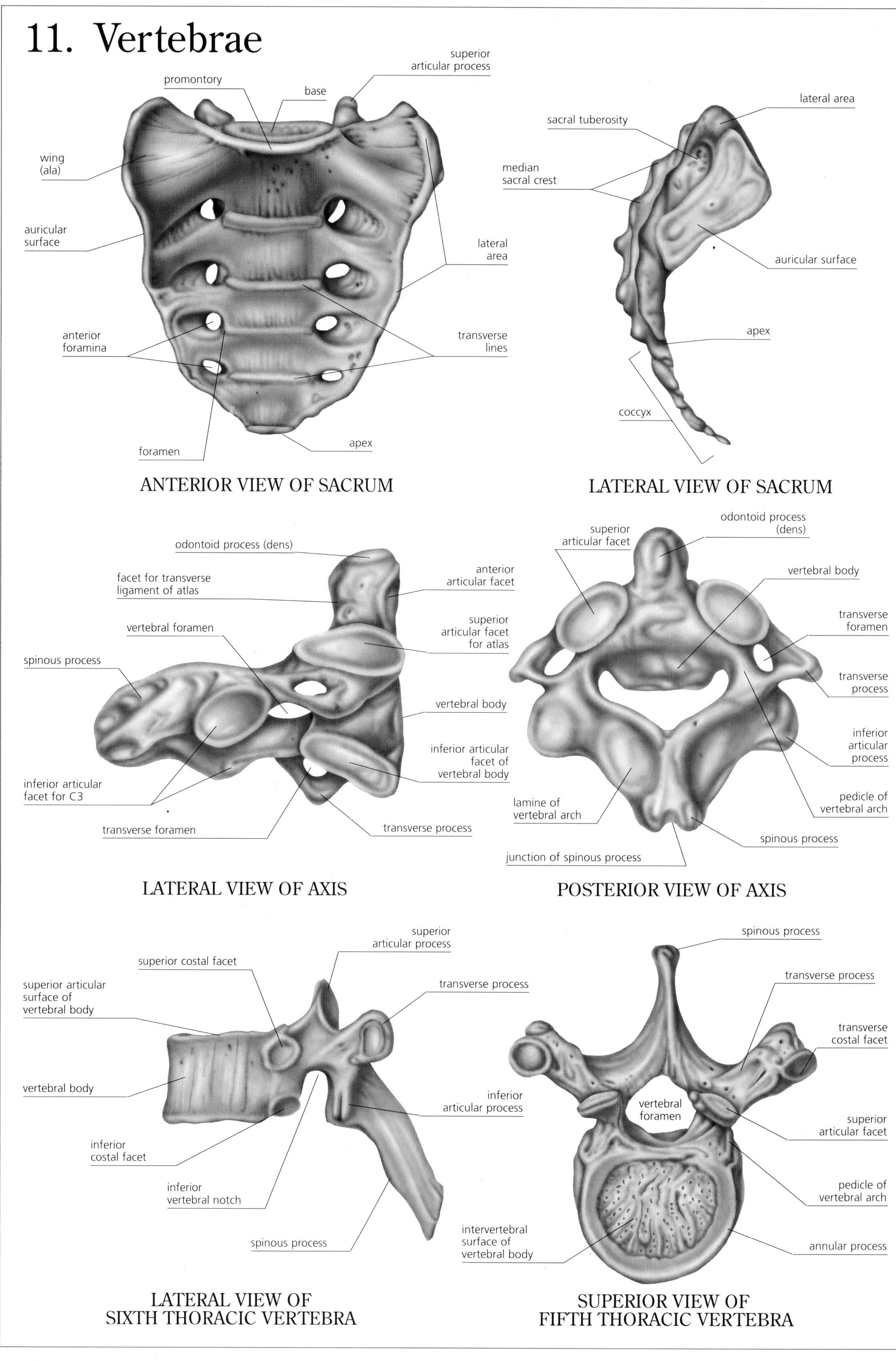

ANTERIOR VIEW OF SACRUM

- superior articular process
- base
- promontory
- wing (ala)
- auricular surface
- anterior foramina
- foramen
- apex
- lateral area
- transverse lines

LATERAL VIEW OF SACRUM

- lateral area
- sacral tuberosity
- median sacral crest
- auricular surface
- apex
- coccyx

LATERAL VIEW OF AXIS

- odontoid process (dens)
- facet for transverse ligament of atlas
- vertebral foramen
- spinous process
- inferior articular facet for C3
- transverse foramen
- anterior articular facet
- superior articular facet for atlas
- vertebral body
- inferior articular facet of vertebral body
- transverse process

POSTERIOR VIEW OF AXIS

- odontoid process (dens)
- superior articular facet
- vertebral body
- transverse foramen
- transverse process
- inferior articular process
- pedicle of vertebral arch
- spinous process
- junction of spinous process
- lamine of vertebral arch

LATERAL VIEW OF
SIXTH THORACIC VERTEBRA

- superior articular process
- superior costal facet
- transverse process
- superior articular surface of vertebral body
- vertebral body
- inferior costal facet
- inferior vertebral notch
- spinous process
- inferior articular process
- intervertebral surface of vertebral body

SUPERIOR VIEW OF
FIFTH THORACIC VERTEBRA

- spinous process
- transverse process
- transverse costal facet
- vertebral foramen
- superior articular facet
- pedicle of vertebral arch
- annular process

12. Vertebrae

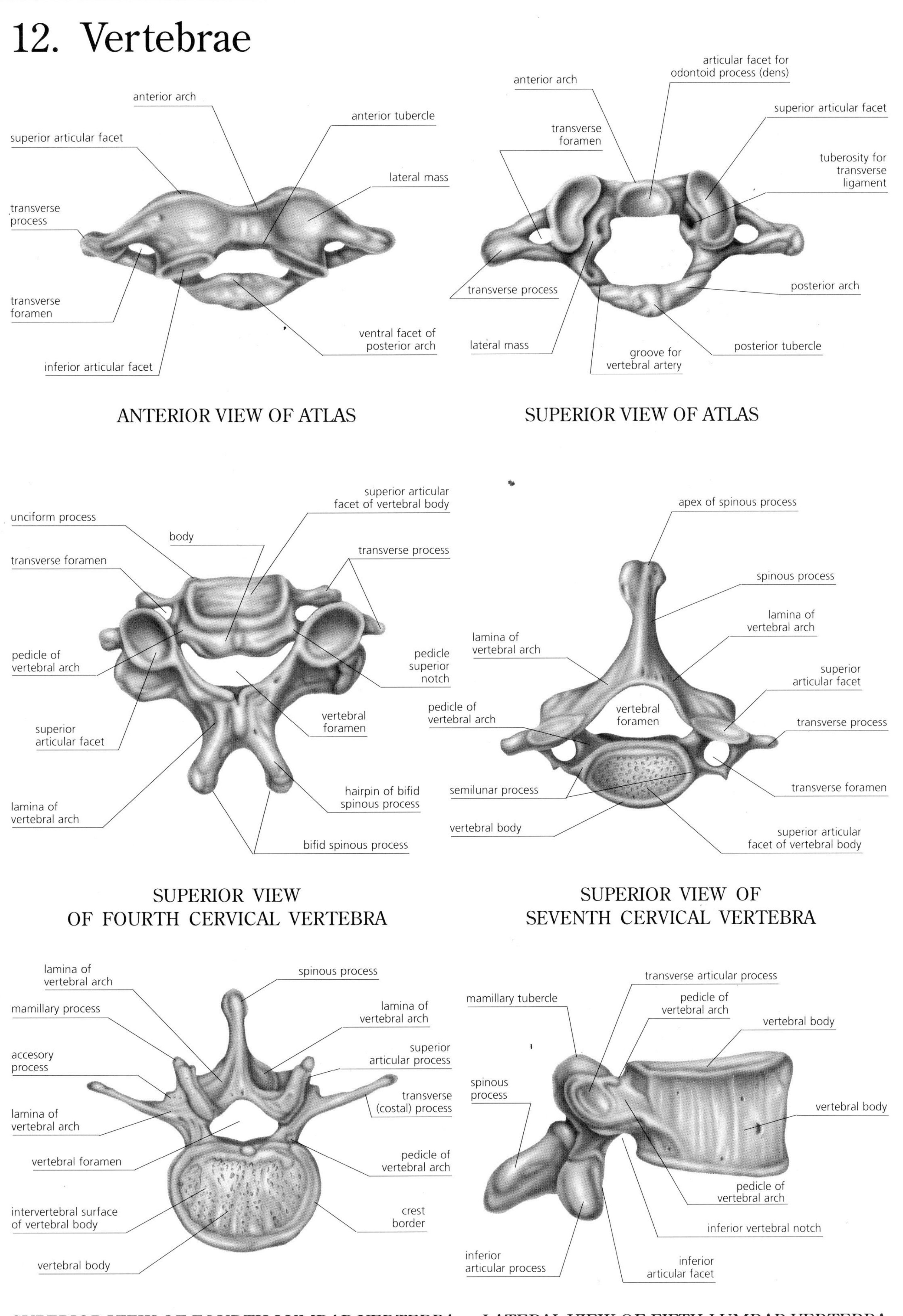

ANTERIOR VIEW OF ATLAS

anterior arch
anterior tubercle
superior articular facet
lateral mass
transverse process
transverse foramen
inferior articular facet
ventral facet of posterior arch

SUPERIOR VIEW OF ATLAS

anterior arch
articular facet for odontoid process (dens)
transverse foramen
superior articular facet
tuberosity for transverse ligament
transverse process
lateral mass
groove for vertebral artery
posterior arch
posterior tubercle

SUPERIOR VIEW OF FOURTH CERVICAL VERTEBRA

unciform process
body
superior articular facet of vertebral body
transverse foramen
transverse process
pedicle of vertebral arch
pedicle superior notch
superior articular facet
vertebral foramen
lamina of vertebral arch
hairpin of bifid spinous process
bifid spinous process

SUPERIOR VIEW OF SEVENTH CERVICAL VERTEBRA

apex of spinous process
spinous process
lamina of vertebral arch
lamina of vertebral arch
superior articular facet
pedicle of vertebral arch
vertebral foramen
transverse process
semilunar process
transverse foramen
vertebral body
superior articular facet of vertebral body

SUPERIOR VIEW OF FOURTH LUMBAR VERTEBRA

lamina of vertebral arch
spinous process
mamillary process
lamina of vertebral arch
accesory process
superior articular process
lamina of vertebral arch
transverse (costal) process
vertebral foramen
pedicle of vertebral arch
intervertebral surface of vertebral body
crest border
vertebral body

LATERAL VIEW OF FIFTH LUMBAR VERTEBRA

transverse articular process
mamillary tubercle
pedicle of vertebral arch
vertebral body
spinous process
vertebral body
inferior articular process
pedicle of vertebral arch
inferior vertebral notch
inferior articular facet

The Muscular System

orbital part of orbicularis oculi muscle
levator labii superioris alaeque nasi muscle
zygomaticus minus muscle
zygomaticus major muscle
masseter muscle
levator labii superioris muscle
buccinator muscle
deppresor anguli oris
levator menti
sternocleidomastoid muscle
trapezius muscle
sternohyoid muscle

triceps brachii muscle
serratus anterior
biceps brachii muscle
rectus abdominis
brachialis muscle
external abdominal oblique
brachioradialis muscle
supinator muscle
flexor digitorum profundus muscle
flexor cappi ulnaris
gluteus medius
anterior superior iliac spine
flexor pollicis longus muscle
tensor fasciae latae muscle
flexor digitorum superficialis muscle
abductor pollicis longus muscle
iliacus muscle
sartorius muscle
psoas major
pectineus
adductor longus
rectus femoris
adductor longus
adductor magnus
vastus lateralis
vastus medialis
rectus femoris tendon
patellar ligament

1. Muscles of the Human Body. Anterior View

peroneus longus muscle
extensor digitorum longus muscle
tibialis anterior
peroneus brevis muscle
interosseous muscles

galea aponeurotica
frontal belly of frontalis muscle
auricularis superior muscle
corrugator supercilii muscle
auricularis anterior muscle
nasalis muscle
depressor anguli oris muscle
orbicularis oris muscle
risorius muscle
mentalis muscle
omohyoid muscle
triangle of neck
deltoid muscles
pectoralis major muscle
coracobrachialis muscle
triceps brachii muscle
biceps brachii muscle
brachialis muscle
triceps brachii muscle
brachialis muscle
humeral epitrochlea
bicipital aponeurosis
supinator muscle
biceps brachii tendon
brachioradialis
extensor carpi radialis longus
pronator teres muscle
flexor carpi radialis
palmaris longus
flexor carpi ulnaris
flexor digitorum superficialis muscle
flexor retinaculum
palmar aponeurosis
abductor digit minimi muscle
dorsal interosseous muscle
flexor digitorum superficialis muscle
fibrous sheath over fingers

crural arch

sartorius muscle

tensor fasciae latae muscle
pectineus muscle
adductor brevis muscle
adductor longus muscle
adductor magnus muscle
rectus femoris muscle
vastus lateralis muscle
vastus medialis muscle
iliotibial ligament
patella
head of fibula
sartorius tendon
gastrocnemius muscle
tibialis anterior muscle

soleus muscle

extensor digitorum longus muscle
peroneus longus muscle
peroneus brevis muscle
extensor hallucis longus muscle
superior extensor retinaculum
extensor digitorum tendons
peroneus tertius tendon

2. Muscles of the Human Body. Posterior View

auricularis superior muscle

occipitalis muscle

auricularis posterior muscle

sternocleidomastoid muscle

splenius muscle

triangle of the neck

trapezius muscle

deltoid muscle

infraspinatus muscle

rhomboideus major muscle

teres major muscle

triceps brachii muscle

epitrochlea of humerus

olecranon

brachialis muscle

brachioradialis

epicondyle of humerus

lumbodorsal fascia

aconeus muscle

internal abdominal oblique

extensor carpi ulnaris

extensor carpi radialis

extensor digtorum communis muscle

external abdominal oblique

abductor pollicis longus

extensor pollicis brevis

extensor digiti minimi

extensor digitorum communis muscle

dorsal interosseous muscle

greater trochanter

gluteus maximus muscle

iliotibial tract

semitendinosus

vastus lateralis muscle

biceps femoris muscle

semimembranous muscle

gracilis

biceps femoris muscle

semitendinous muscle

semimembranous muscle

plantaris muscle

biceps femoris muscle

lateral head of gastrocnemius muscle

medial head of gastrocnemius muscle

raphe

gastrocnemius muscle

soleus muscle

gastrocnemius tendon

soleus muscle

peroneus longus muscle

plantaris tendon

peroneus brevis muscle

flexor longus pollicic

flexor digitorum longus

Achilles tendon

peroneal retinaculum

peroneus brevis tendon

peroneus longus tendon

latissimus dorsi muscle

cubital posterior

sartorius muscles

galea aponeurotica

transversalis cervicis

vertebra prominens

spine of scapula

iliocostalis muscle

latissimus dorsii muscle

triceps brachii muscle

tendon of triceps brachii muscle

brachialis muscle

triceps brachii muscle

brachioradialis muscle

iliac crest

lumbodorsal fascia

gluteus medius

tensor fasciae latae muscle

extensor pollicis brevis muscle

extensor indicis muscle

extensor retinaculum

extensor pollicis longus muscle

extensor digitorum communis muscle

symphysis

adductor magnus muscle

gracilis muscle

vastus lateralis muscle

semitendonosus

biceps femoris muscle

vastus lateralis muscle

sartorius muscles

biceps femoris muscle

lateral head of gastrocnemius muscle

raphe

gastrocnemius muscles

gastrocnemius muscles

soleus muscle

peroneus longus muscle

plantaris tendon

peroneus brevis muscle

flexor digitorum longus muscle

flexor retinaculum

medial malleolus

lateral malleolus

superior peroneal retinaculum

inferior peroneal retinaculum

extensor communis digitorum

tuberosity of calcaneus

occipitalis muscle

temporalis

masseter muscle

auricularis posterior muscle

buccinator muscle

splenius muscle

risorius muscle

serratus muscle

trapezius muscle

levator scapulae muscle

spine of scapula

deltoid muscle

infraspinatus fascia

infraspinatus muscle

teres minor

teres major

long head of triceps brachii muscle

short head of triceps brachii muscle

biceps brachii muscle

brachialis muscle

tendon of triceps brachii muscle

medial head of triceps brachii muscle

medial intermuscular septem

lateral epicondyle

extensor carpi radialis longus muscle

olecranon

anconeus muscle

extensor digitorum muscle

radius

gluteus maximus muscle

tendons of extensor digitorum muscle

opponens pollicis longus muscle

extensor retinaculum

long head of biceps femoris muscle

iliotibial tract

short head of biceps femoris muscle

semimembranous muscle

head of fibula

plantaris muscle

gastroncnemius muscle

frontalis

orbicularis oculi muscle

zygomaticus

orbicularis oris muscle

omohyoid muscle

sternocleidomastoid muscle

inferior belly of omohyoid muscle

acromion

clavicle

pectoralis major muscle

serratus anterior muscles

intercostal muscle

pronator teres

external abdominal oblique muscle

brachioradialis muscle

extensor carpi radialis brevis

rectus sheath

external oblique aponeurosis

sartorius

flexor pollicis longus muscle

extensor pollicis brevis muscle

flexor pollicis longus tendon

extensor brevis tendon

extensor pollicis longus tendon

abductor pollicis

dorsal interosseous muscles

iliotibial tract

rectus femoris muscle

vastus lateralis muscle

vastus medialis muscle

iliotibial tract

patella

head of fibula

patellar ligament

tibial tuberosity

peroneus longus muscle

tibialis anterior muscle

anterior intermuscular septum

peroneus longus muscle

extensor digitorum longus muscle

tibialis anterior tendon

extensor digitorum longus

inferior extensor retinaculum

extensor digitorum brevis

peroneus tertius tendon

extensor hallucis brevis muscle

extensor digitorum longus tendons

3. Muscles of the Human Body. Lateral View

soleus muscle

Achilles (calcaneal) tendon

lateral malleolus

peroneal retinaculum

calcaneal tuberosity

peroneus brevis tendon

4. Muscles of the Head

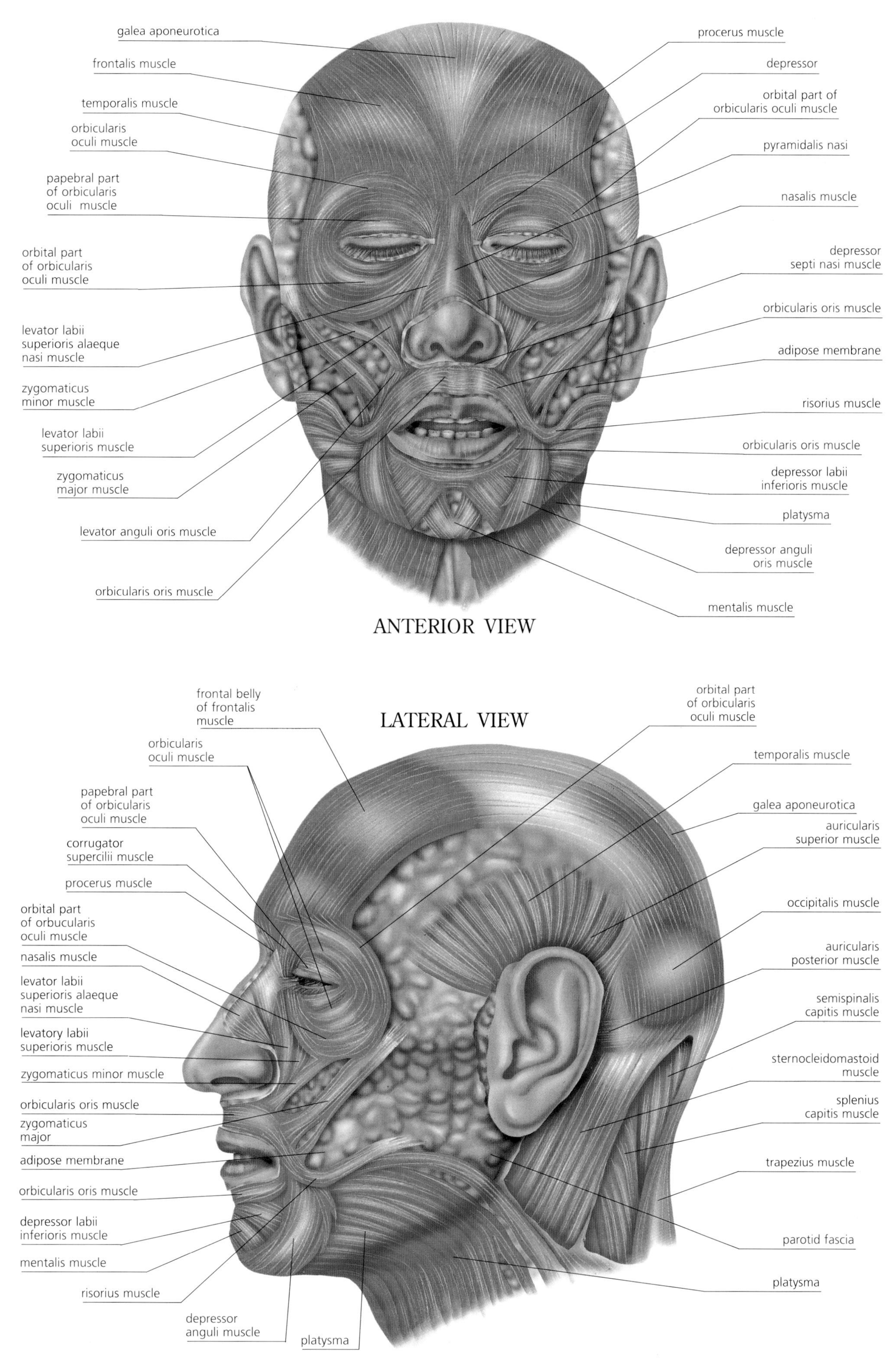

galea aponeurotica

frontalis muscle

temporalis muscle

orbicularis oculi muscle

papebral part of orbicularis oculi muscle

orbital part of orbicularis oculi muscle

levator labii superioris alaeque nasi muscle

zygomaticus minor muscle

levator labii superioris muscle

zygomaticus major muscle

levator anguli oris muscle

orbicularis oris muscle

procerus muscle

depressor

orbital part of orbicularis oculi muscle

pyramidalis nasi

nasalis muscle

depressor septi nasi muscle

orbicularis oris muscle

adipose membrane

risorius muscle

orbicularis oris muscle

depressor labii inferioris muscle

platysma

depressor anguli oris muscle

mentalis muscle

ANTERIOR VIEW

LATERAL VIEW

frontal belly of frontalis muscle

orbicularis oculi muscle

papebral part of orbicularis oculi muscle

corrugator supercilii muscle

procerus muscle

orbital part of orbucularis oculi muscle

nasalis muscle

levator labii superioris alaeque nasi muscle

levatory labii superioris muscle

zygomaticus minor muscle

orbicularis oris muscle

zygomaticus major

adipose membrane

orbicularis oris muscle

depressor labii inferioris muscle

mentalis muscle

risorius muscle

depressor anguli muscle

platysma

orbital part of orbicularis oculi muscle

temporalis muscle

galea aponeurotica

auricularis superior muscle

occipitalis muscle

auricularis posterior muscle

semispinalis capitis muscle

sternocleidomastoid muscle

splenius capitis muscle

trapezius muscle

parotid fascia

platysma

5. Hands and Feet

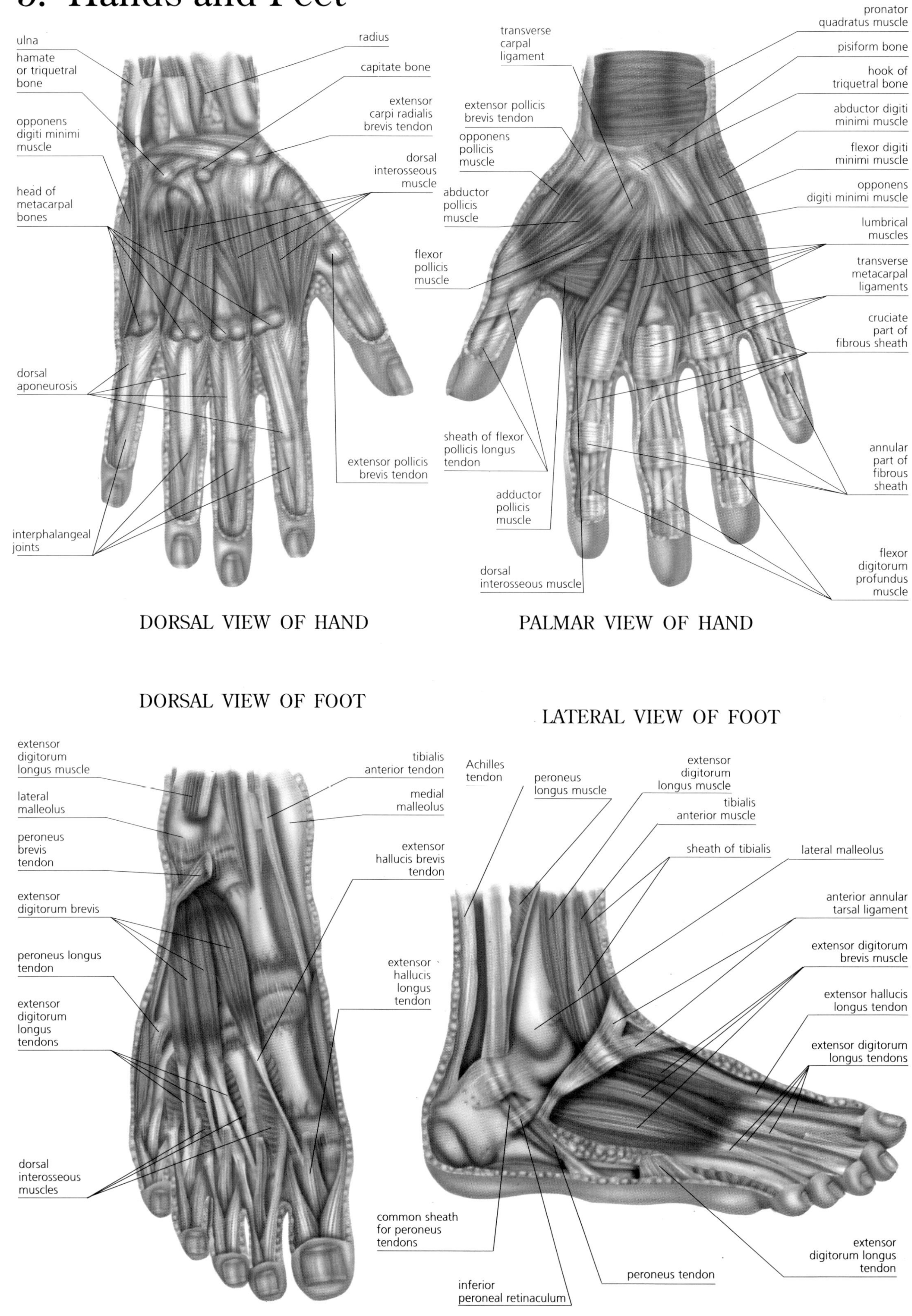

ulna
hamate or triquetral bone
opponens digiti minimi muscle
head of metacarpal bones
dorsal aponeurosis
interphalangeal joints

radius
capitate bone
extensor carpi radialis brevis tendon
dorsal interosseous muscle
extensor pollicis brevis tendon

transverse carpal ligament
extensor pollicis brevis tendon
opponens pollicis muscle
abductor pollicis muscle
flexor pollicis muscle
sheath of flexor pollicis longus tendon
adductor pollicis muscle
dorsal interosseous muscle

pronator quadratus muscle
pisiform bone
hook of triquetral bone
abductor digiti minimi muscle
flexor digiti minimi muscle
opponens digiti minimi muscle
lumbrical muscles
transverse metacarpal ligaments
cruciate part of fibrous sheath
annular part of fibrous sheath
flexor digitorum profundus muscle

DORSAL VIEW OF HAND

PALMAR VIEW OF HAND

DORSAL VIEW OF FOOT

LATERAL VIEW OF FOOT

extensor digitorum longus muscle
lateral malleolus
peroneus brevis tendon
extensor digitorum brevis
peroneus longus tendon
extensor digitorum longus tendons
dorsal interosseous muscles

tibialis anterior tendon
medial malleolus
extensor hallucis brevis tendon
extensor hallucis longus tendon

Achilles tendon
peroneus longus muscle
common sheath for peroneus tendons
inferior peroneal retinaculum
peroneus tendon

extensor digitorum longus muscle
tibialis anterior muscle
sheath of tibialis
lateral malleolus
anterior annular tarsal ligament
extensor digitorum brevis muscle
extensor hallucis longus tendon
extensor digitorum longus tendons
extensor digitorum longus tendon

6. Shapes and Types of Muscles and Muscular Insertions

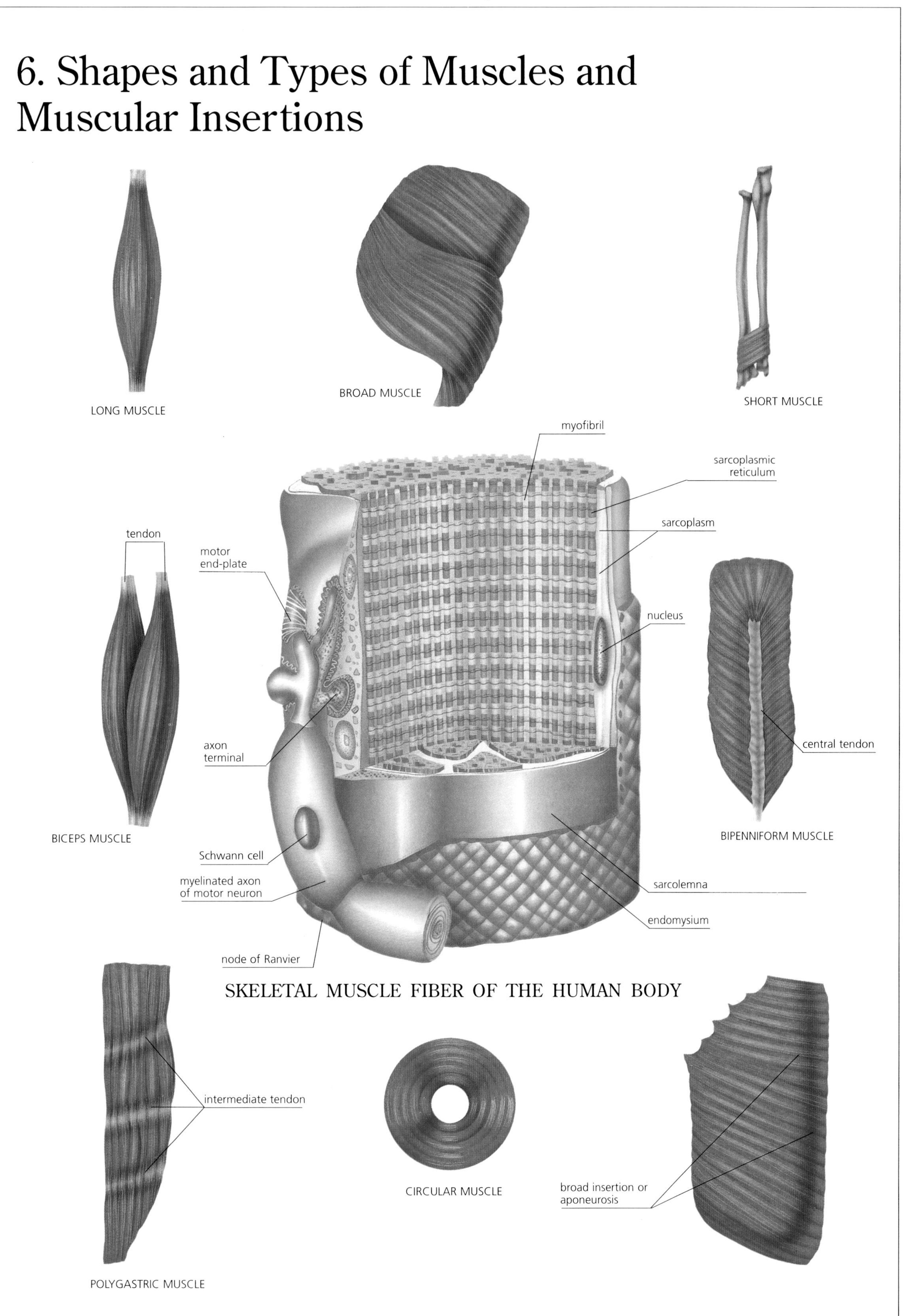

LONG MUSCLE

BROAD MUSCLE

SHORT MUSCLE

myofibril

sarcoplasmic reticulum

sarcoplasm

tendon

motor end-plate

nucleus

axon terminal

central tendon

BICEPS MUSCLE

BIPENNIFORM MUSCLE

Schwann cell

myelinated axon of motor neuron

sarcolemna

endomysium

node of Ranvier

SKELETAL MUSCLE FIBER OF THE HUMAN BODY

intermediate tendon

CIRCULAR MUSCLE

broad insertion or aponeurosis

POLYGASTRIC MUSCLE

7. Movements of the Forearm

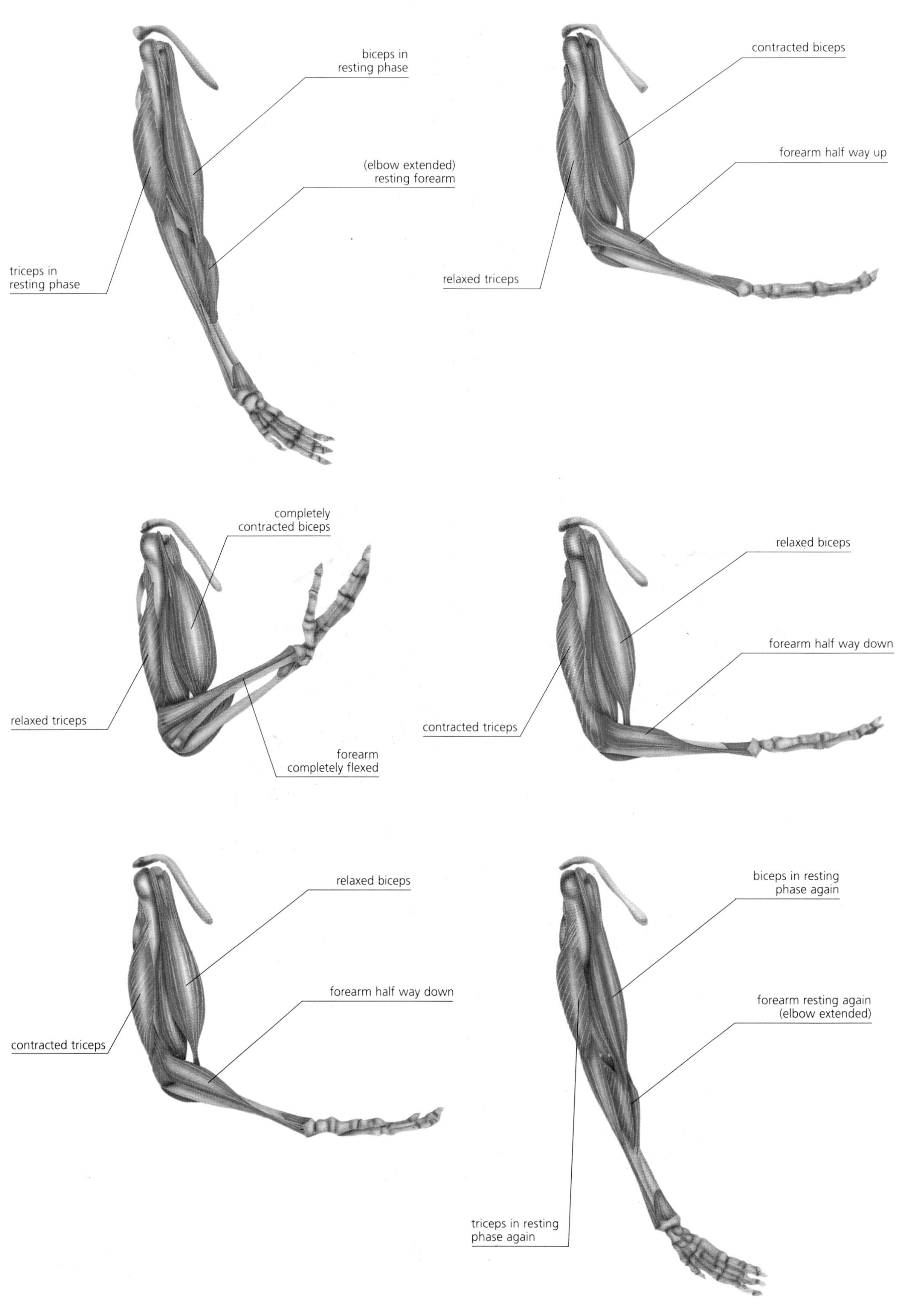

biceps in
resting phase

(elbow extended)
resting forearm

triceps in
resting phase

contracted biceps

forearm half way up

relaxed triceps

completely
contracted biceps

relaxed triceps

forearm
completely flexed

relaxed biceps

forearm half way down

contracted triceps

relaxed biceps

forearm half way down

contracted triceps

biceps in resting
phase again

forearm resting again
(elbow extended)

triceps in resting
phase again

The Digestive System

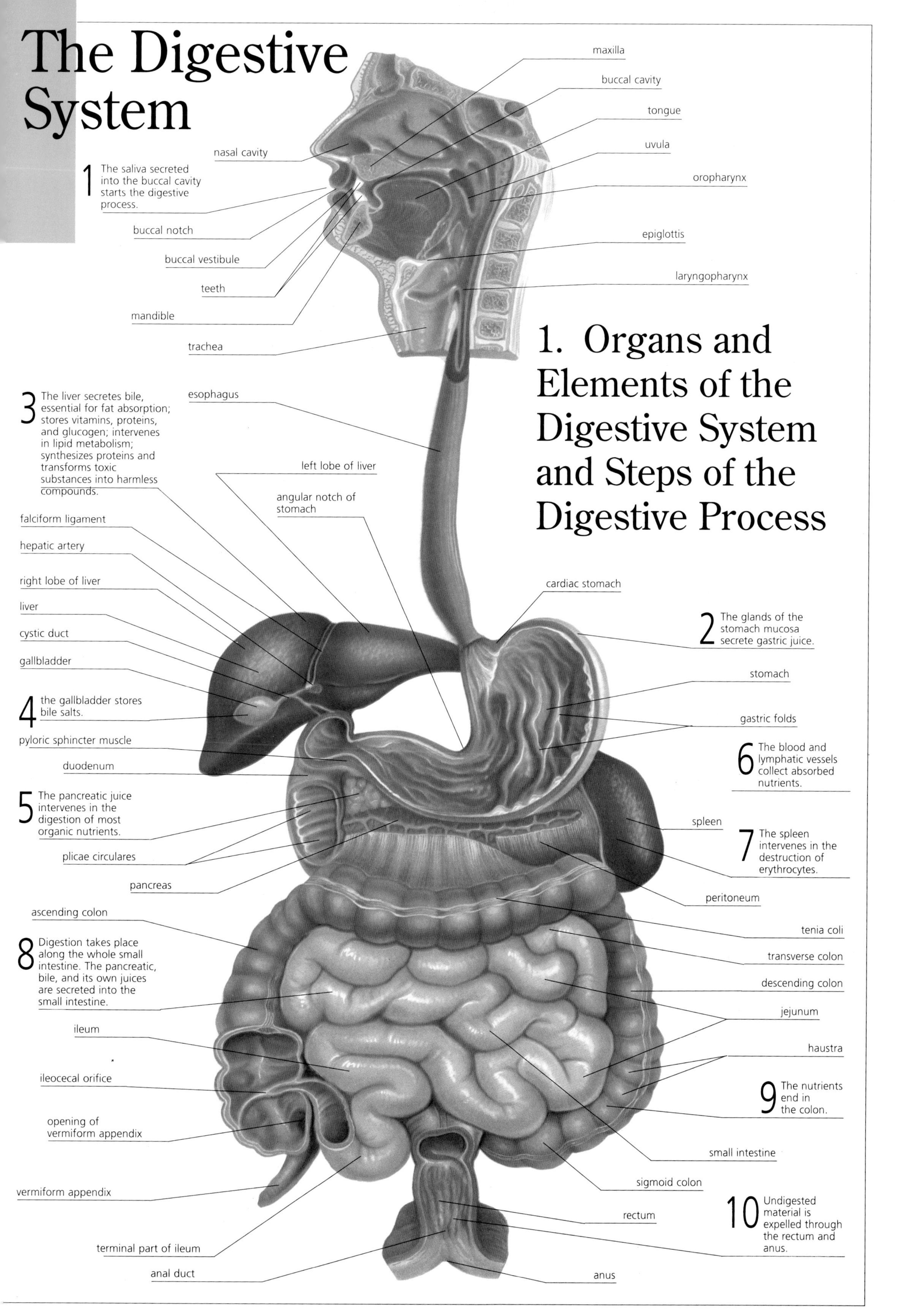

1. Organs and Elements of the Digestive System and Steps of the Digestive Process

1 The saliva secreted into the buccal cavity starts the digestive process.

2 The glands of the stomach mucosa secrete gastric juice.

3 The liver secretes bile, essential for fat absorption; stores vitamins, proteins, and glucogen; intervenes in lipid metabolism; synthesizes proteins and transforms toxic substances into harmless compounds.

4 the gallbladder stores bile salts.

5 The pancreatic juice intervenes in the digestion of most organic nutrients.

6 The blood and lymphatic vessels collect absorbed nutrients.

7 The spleen intervenes in the destruction of erythrocytes.

8 Digestion takes place along the whole small intestine. The pancreatic, bile, and its own juices are secreted into the small intestine.

9 The nutrients end in the colon.

10 Undigested material is expelled through the rectum and anus.

maxilla
buccal cavity
tongue
uvula
oropharynx
epiglottis
laryngopharynx
nasal cavity
buccal notch
buccal vestibule
teeth
mandible
trachea
esophagus
left lobe of liver
angular notch of stomach
falciform ligament
hepatic artery
right lobe of liver
liver
cystic duct
gallbladder
pyloric sphincter muscle
duodenum
plicae circulares
pancreas
ascending colon
ileum
ileocecal orifice
opening of vermiform appendix
vermiform appendix
terminal part of ileum
anal duct
cardiac stomach
stomach
gastric folds
spleen
peritoneum
tenia coli
transverse colon
descending colon
jejunum
haustra
small intestine
sigmoid colon
rectum
anus

2. Stomach

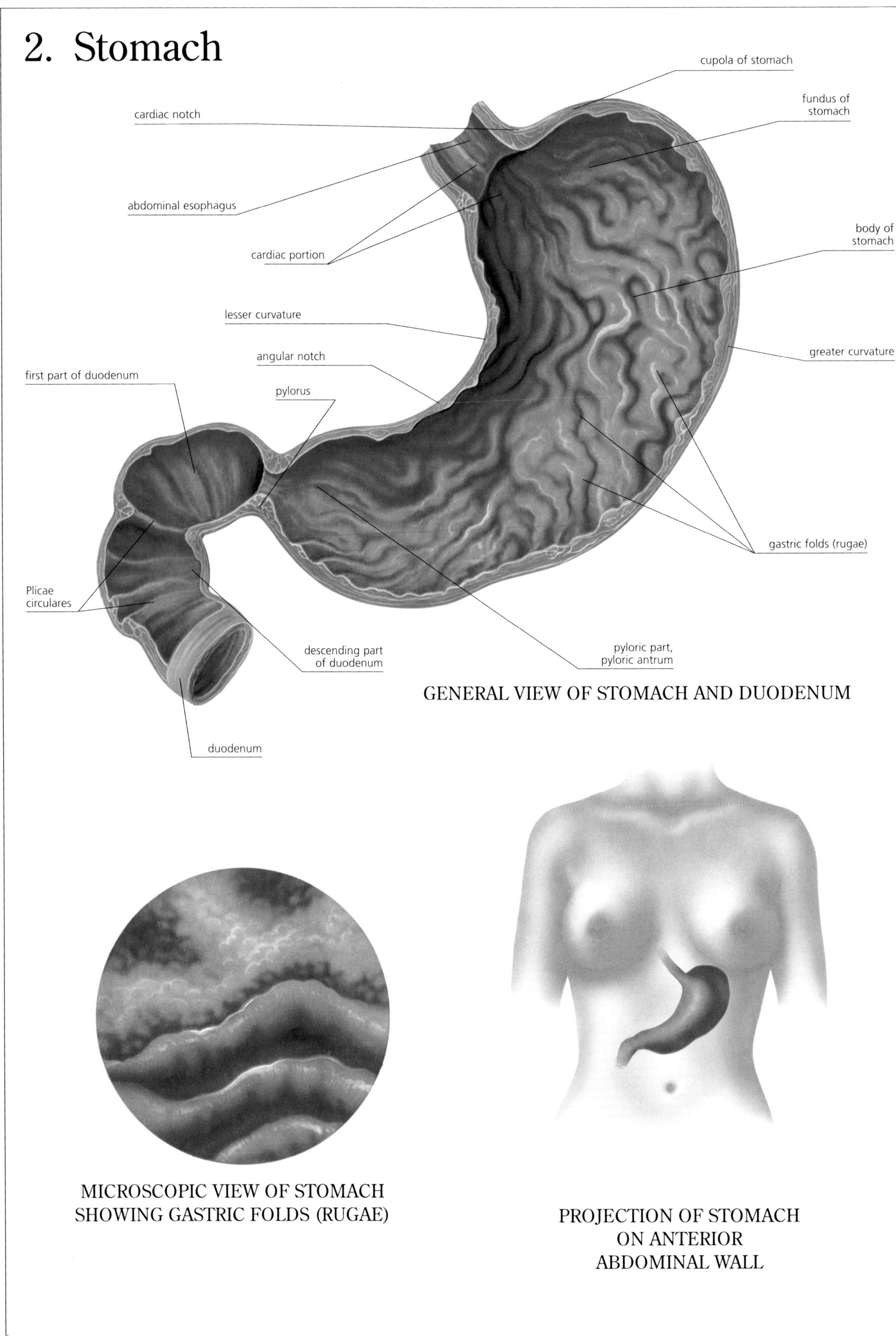

cupola of stomach

fundus of stomach

cardiac notch

abdominal esophagus

body of stomach

cardiac portion

lesser curvature

greater curvature

angular notch

first part of duodenum

pylorus

Plicae circulares

descending part of duodenum

gastric folds (rugae)

pyloric part, pyloric antrum

duodenum

GENERAL VIEW OF STOMACH AND DUODENUM

MICROSCOPIC VIEW OF STOMACH
SHOWING GASTRIC FOLDS (RUGAE)

PROJECTION OF STOMACH
ON ANTERIOR
ABDOMINAL WALL

3. Stomach

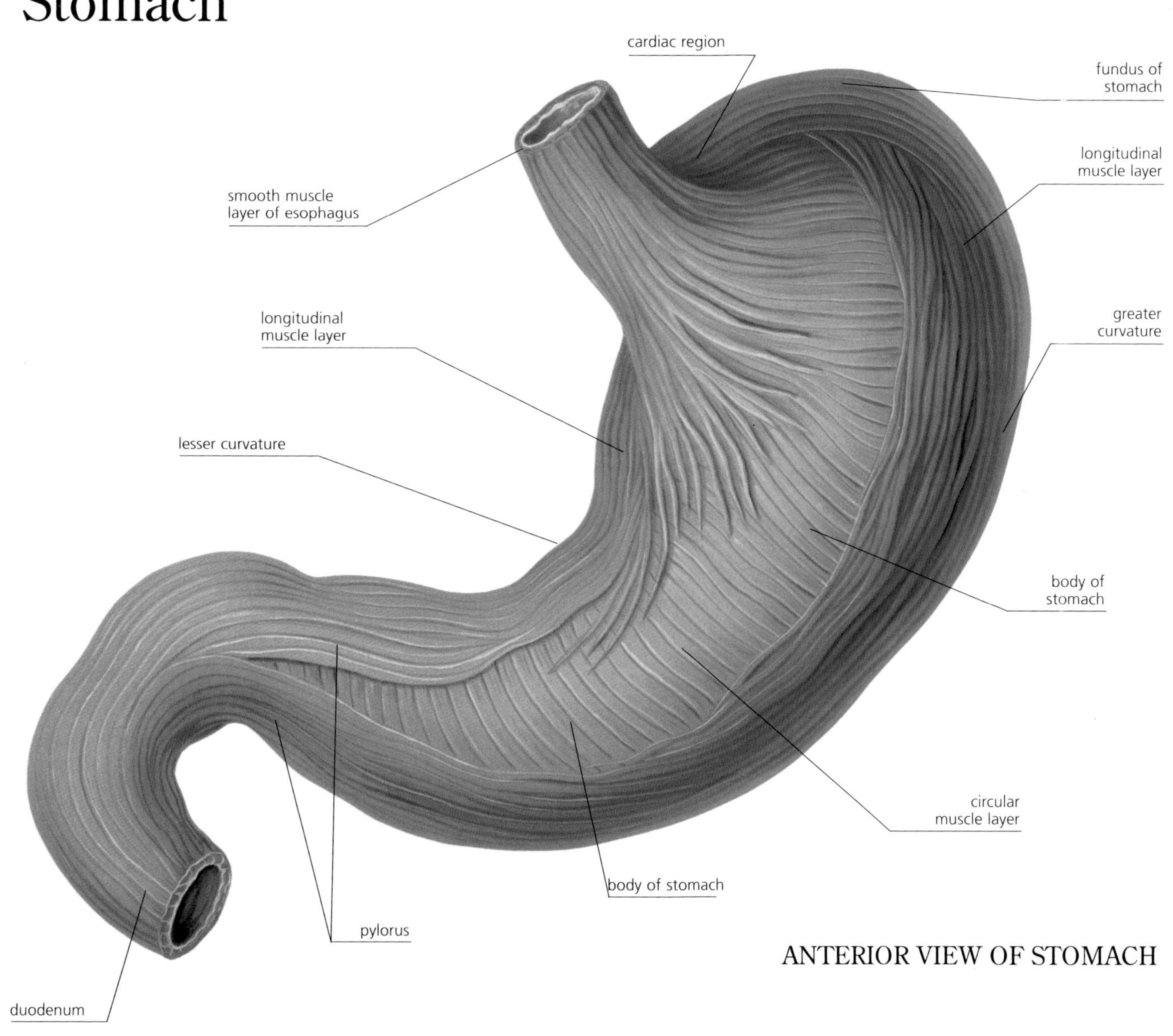

cardiac region

fundus of stomach

longitudinal muscle layer

smooth muscle layer of esophagus

longitudinal muscle layer

greater curvature

lesser curvature

body of stomach

circular muscle layer

body of stomach

pylorus

duodenum

ANTERIOR VIEW OF STOMACH

SCHEMATIC SECTION OF STOMACH WALL

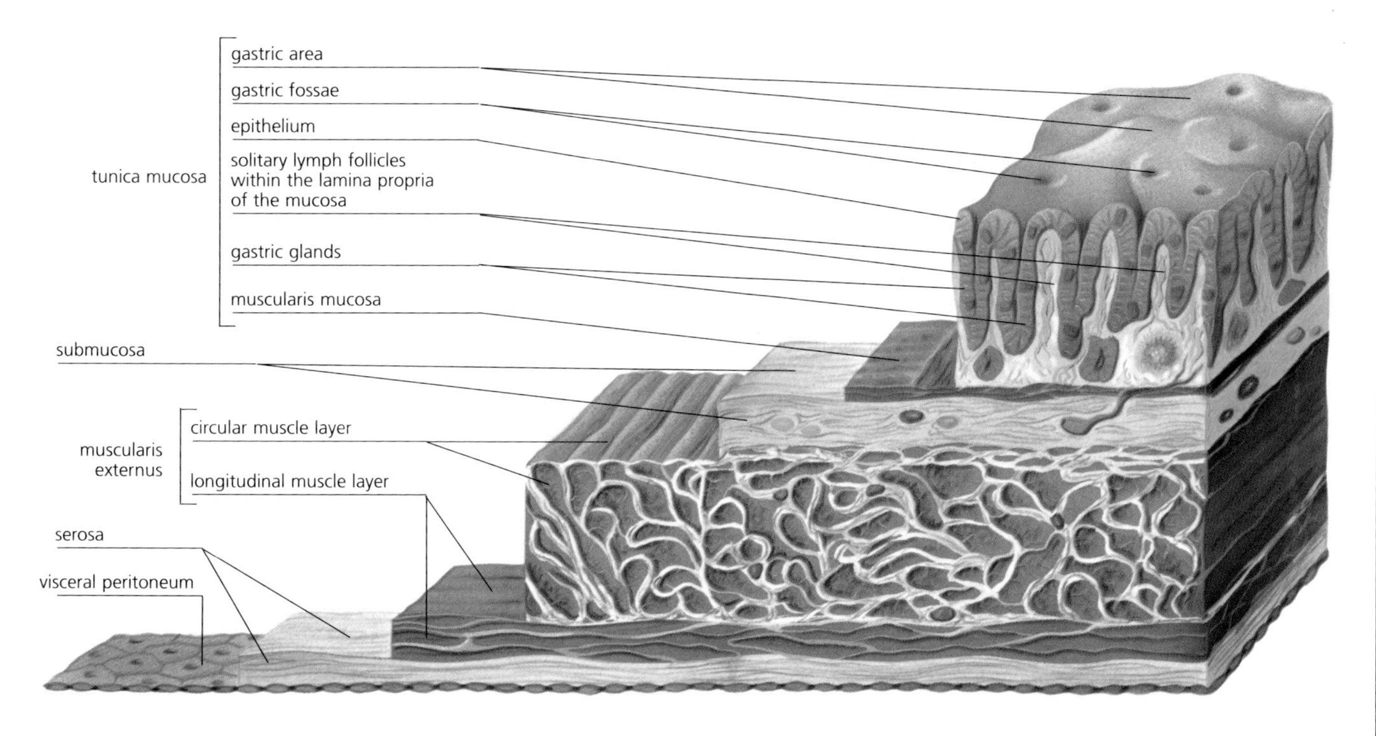

gastric area

gastric fossae

epithelium

tunica mucosa

solitary lymph follicles within the lamina propria of the mucosa

gastric glands

muscularis mucosa

submucosa

muscularis externus

circular muscle layer

longitudinal muscle layer

serosa

visceral peritoneum

4. Duodenum and Pancreas

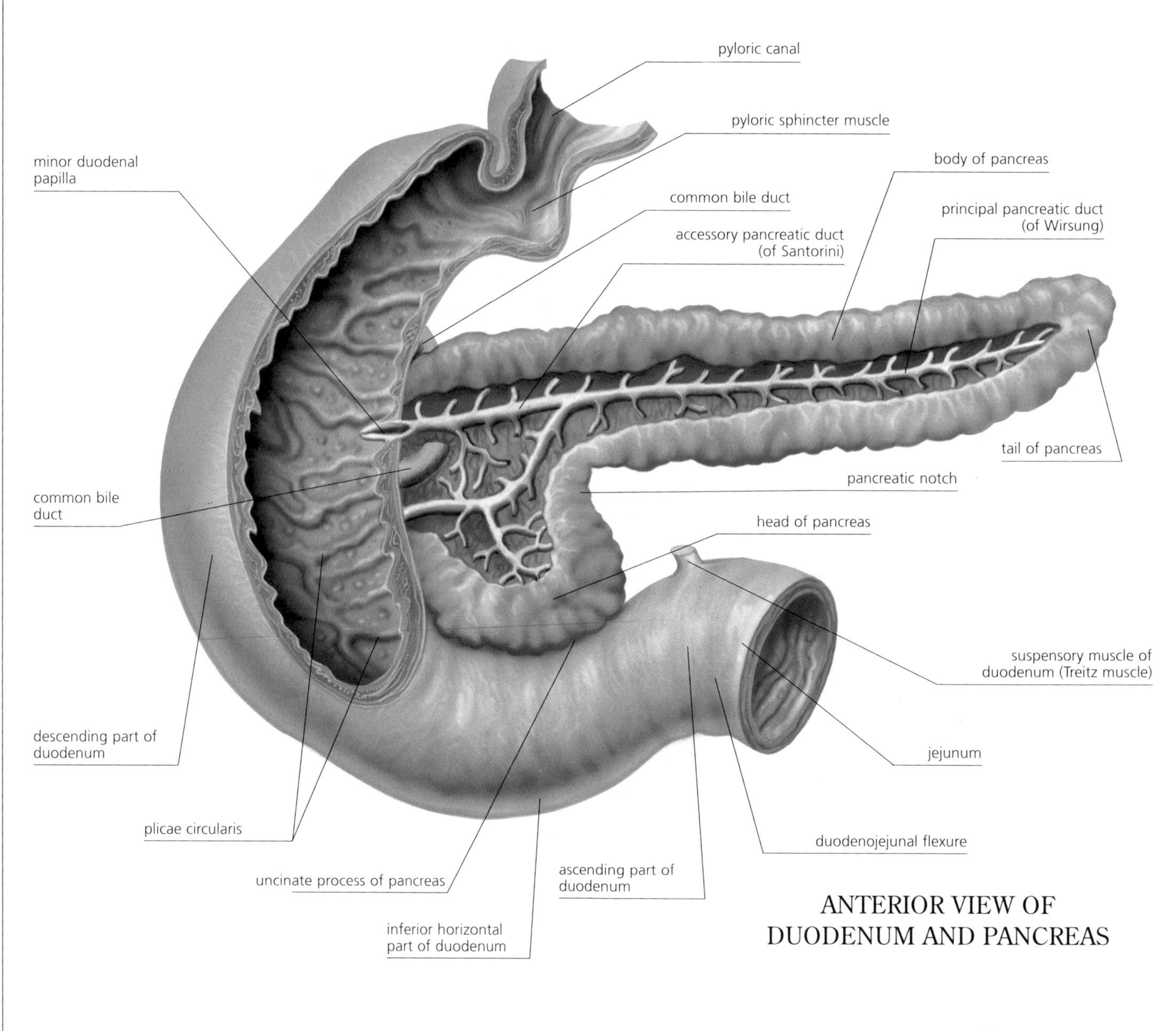

pyloric canal

pyloric sphincter muscle

body of pancreas

minor duodenal papilla

common bile duct

principal pancreatic duct (of Wirsung)

accessory pancreatic duct (of Santorini)

tail of pancreas

pancreatic notch

common bile duct

head of pancreas

suspensory muscle of duodenum (Treitz muscle)

descending part of duodenum

jejunum

plicae circularis

duodenojejunal flexure

uncinate process of pancreas

ascending part of duodenum

inferior horizontal part of duodenum

ANTERIOR VIEW OF DUODENUM AND PANCREAS

MICROSCOPIC VIEW OF ACINAR CELLS OF THE PANCREAS

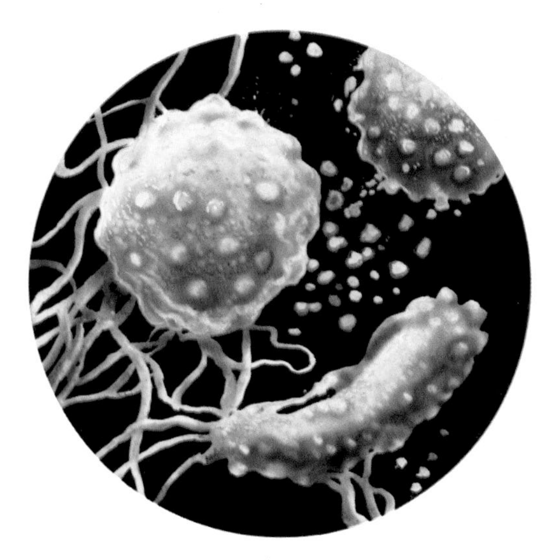

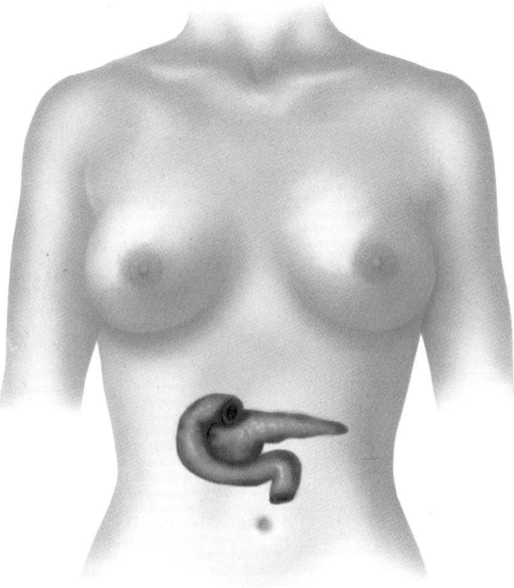

PROJECTION OF DUODENUM AND PANCREAS ON ANTERIOR ABDOMINAL WALL

RETROPERITONEAL ORGANS OF POSTERIOR ABDOMINAL WALL AND EPIGASTRIC VESSELS

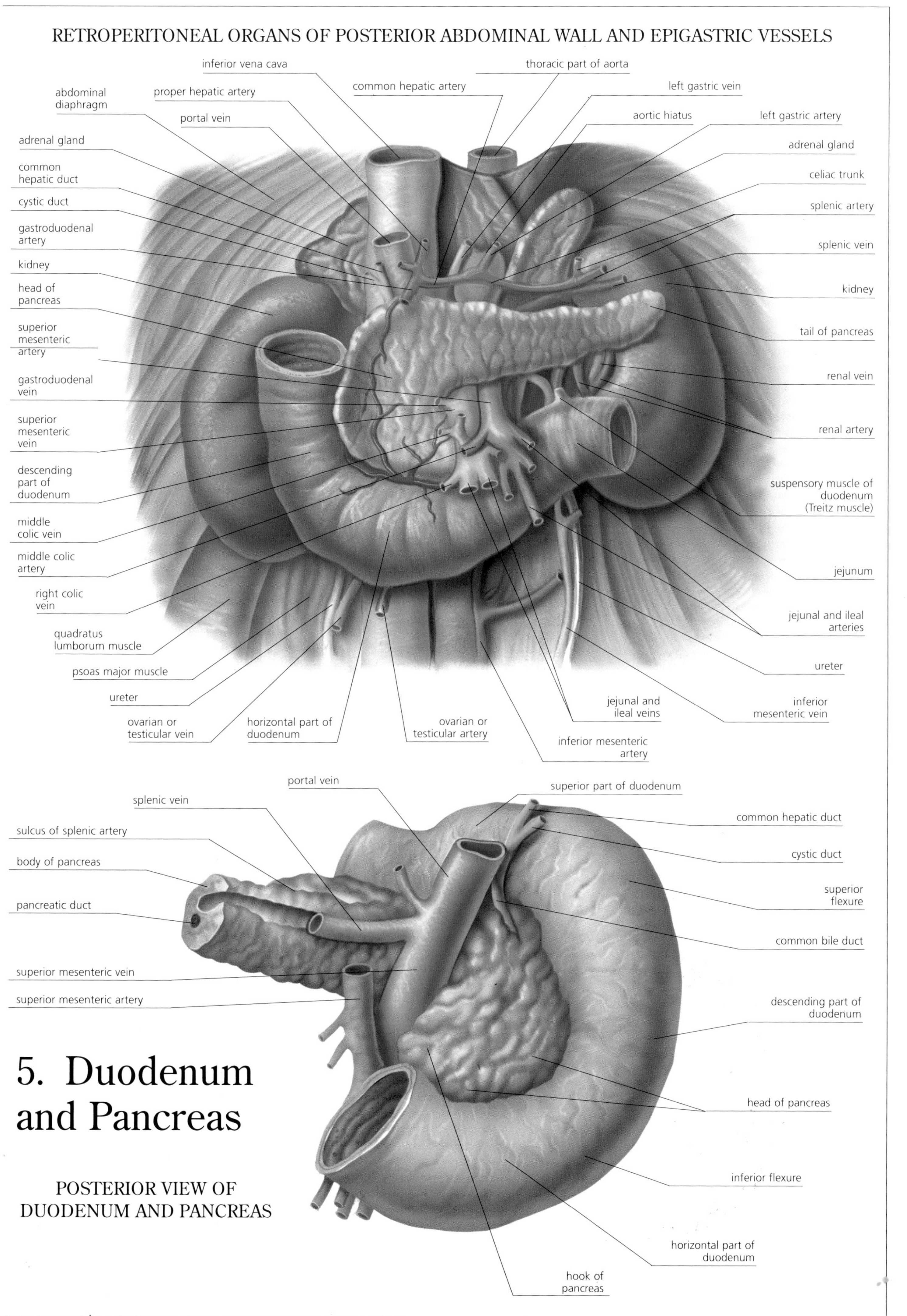

5. Duodenum and Pancreas

POSTERIOR VIEW OF DUODENUM AND PANCREAS

6. Liver

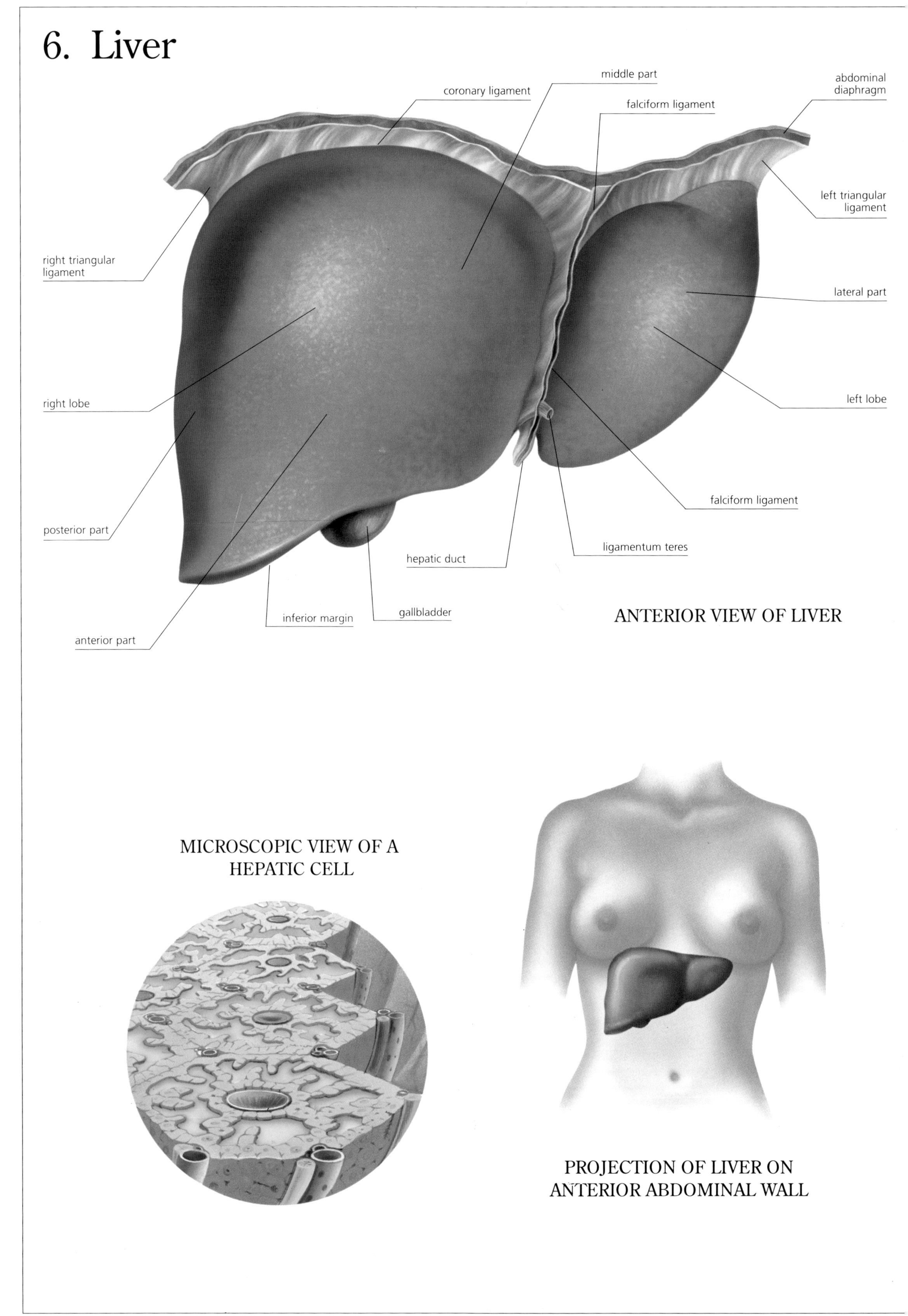

coronary ligament
middle part
falciform ligament
abdominal diaphragm
right triangular ligament
left triangular ligament
lateral part
right lobe
left lobe
posterior part
falciform ligament
ligamentum teres
hepatic duct
inferior margin
gallbladder
anterior part

ANTERIOR VIEW OF LIVER

MICROSCOPIC VIEW OF A HEPATIC CELL

PROJECTION OF LIVER ON ANTERIOR ABDOMINAL WALL

7. Liver and Gallbladder

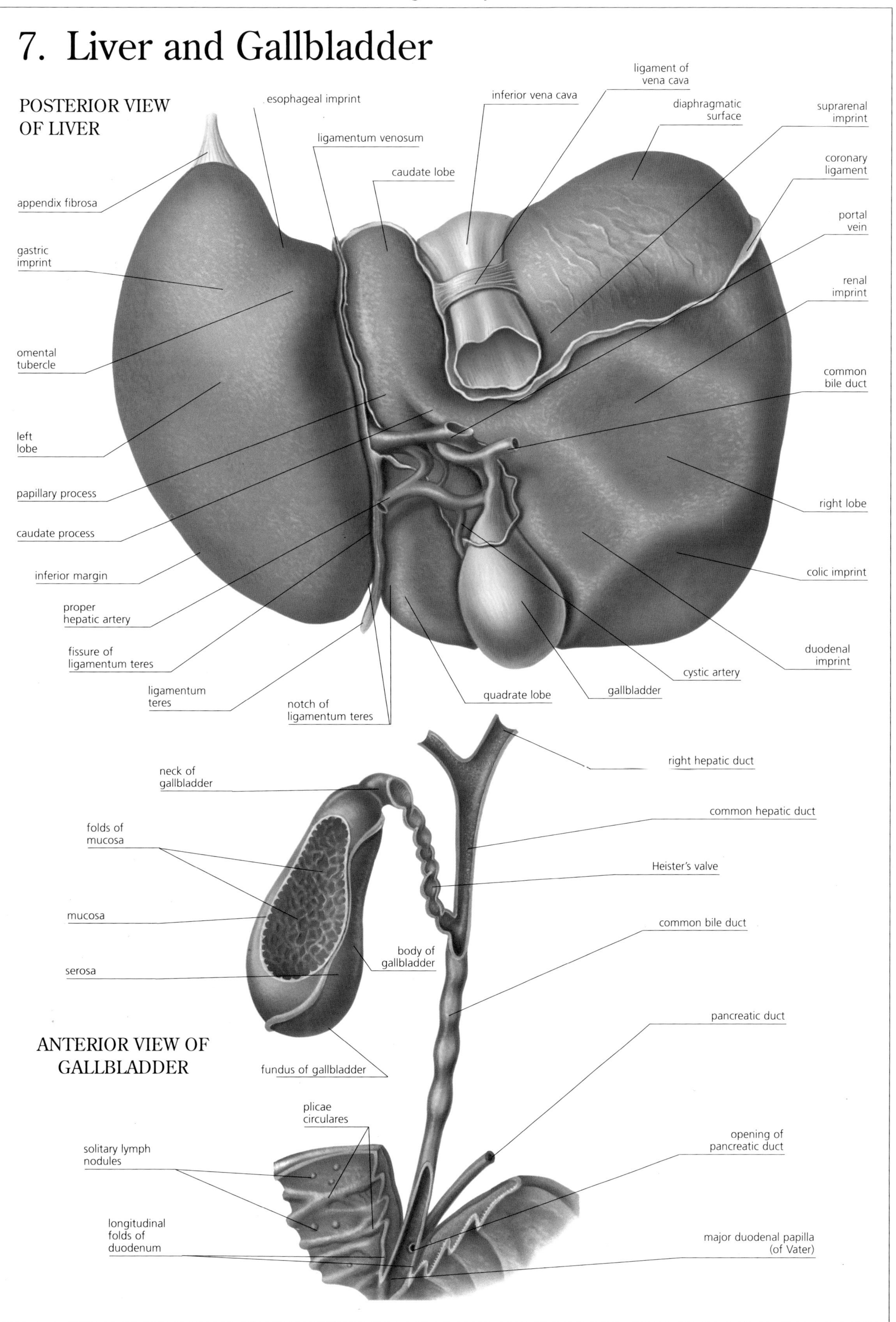

POSTERIOR VIEW
OF LIVER

ligament of
vena cava

esophageal imprint

inferior vena cava

diaphragmatic
surface

suprarenal
imprint

ligamentum venosum

caudate lobe

coronary
ligament

appendix fibrosa

portal
vein

gastric
imprint

renal
imprint

common
bile duct

omental
tubercle

left
lobe

papillary process

right lobe

caudate process

colic imprint

inferior margin

proper
hepatic artery

duodenal
imprint

fissure of
ligamentum teres

cystic artery

gallbladder

ligamentum
teres

notch of
ligamentum teres

quadrate lobe

right hepatic duct

neck of
gallbladder

common hepatic duct

folds of
mucosa

Heister's valve

mucosa

common bile duct

serosa

body of
gallbladder

pancreatic duct

ANTERIOR VIEW OF
GALLBLADDER

fundus of gallbladder

plicae
circulares

opening of
pancreatic duct

solitary lymph
nodules

longitudinal
folds of
duodenum

major duodenal papilla
(of Vater)

8. Small Intestine

ANTERIOR VIEW
OF SMALL
INTESTINE
"SURROUNDED"
BY LARGE
INTESTINE

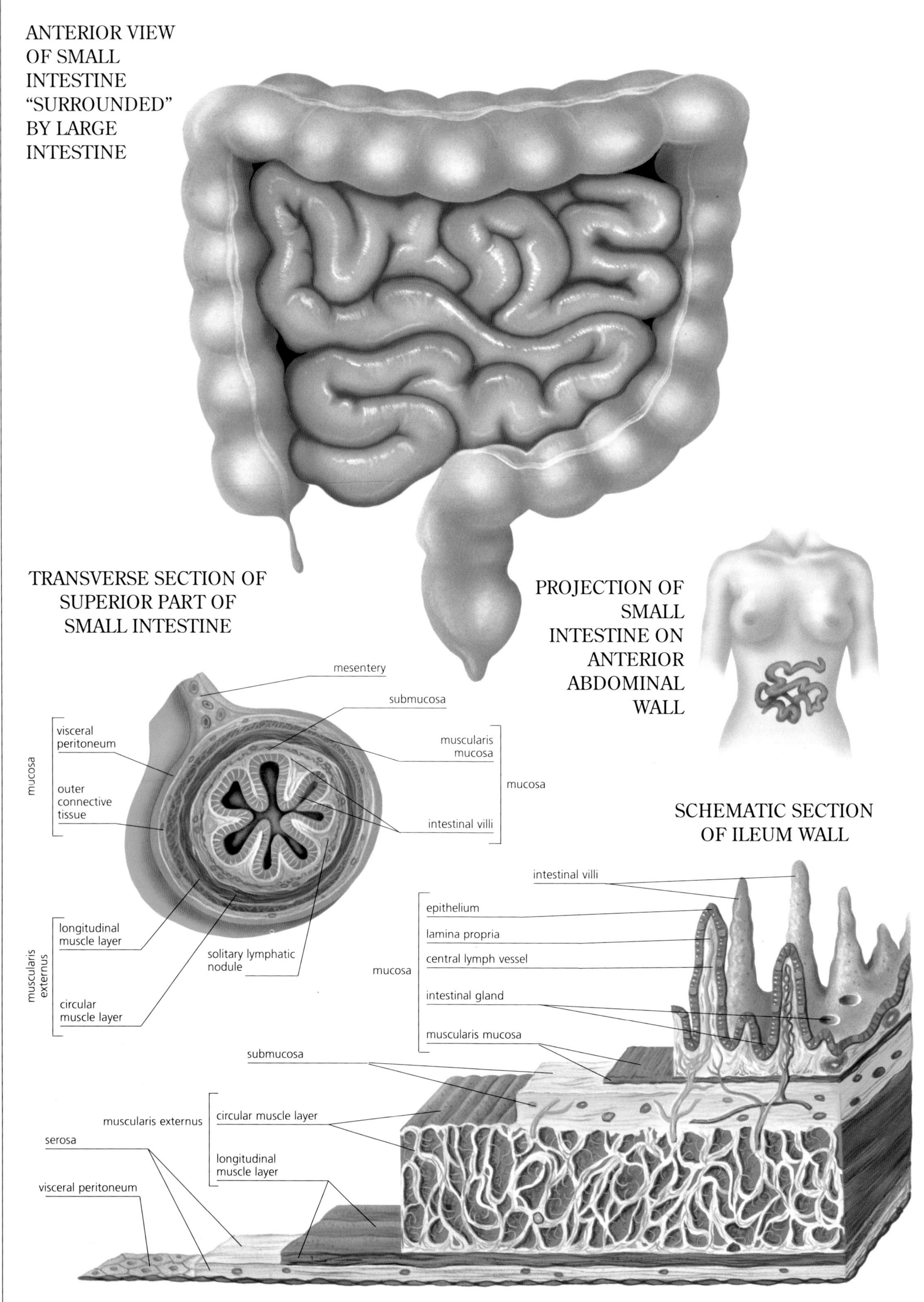

TRANSVERSE SECTION OF
SUPERIOR PART OF
SMALL INTESTINE

PROJECTION OF
SMALL
INTESTINE ON
ANTERIOR
ABDOMINAL
WALL

SCHEMATIC SECTION
OF ILEUM WALL

mesentery

submucosa

muscularis
mucosa

mucosa

intestinal villi

visceral
peritoneum

mucosa

outer
connective
tissue

longitudinal
muscle layer

solitary lymphatic
nodule

muscularis
externus

circular
muscle layer

intestinal villi

epithelium

lamina propria

central lymph vessel

intestinal gland

muscularis mucosa

mucosa

submucosa

muscularis externus

circular muscle layer

serosa

longitudinal
muscle layer

visceral peritoneum

9. Large Intestine

transverse colon

left colic flexure

right colic flexure

tenia coli

ascending colon

paracolic sulcus

cecum

vermiform appendix

haustra

descending colon

ANTERIOR VIEW OF
LARGE INTESTINE

superior ileocecal recess

sigmoid colon

rectal ampulla

rectum

PROJECTION OF
LARGE INTESTINE
ON ANTERIOR
ABDOMINAL WALL

FRONTAL SECTION OF ASCENDING
COLON, CECUM, AND VERMIFORM
APPENDIX

tenia coli

ascending colon

semilunar folds

haustra

ileocecal sphincter

ileocecal sphincter

ileum

solitary lymph nodules

semilunar folds

vermiform appendix

ileocecal sphincter

mucosa

cecum

opening of vermiform appendix

tenia coli

submucosa

SCHEMATIC SECTION
OF COLON WALL

intestinal glands

epithelium

lamina propria

lymphatic nodules

muscularis mucosa

muscularis externus

circular muscle layer

longitudinal muscle layer

visceral peritoneum

serosa

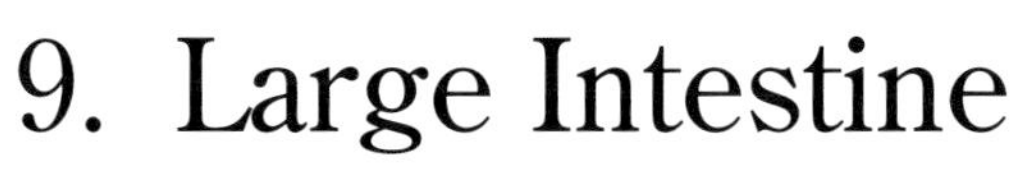

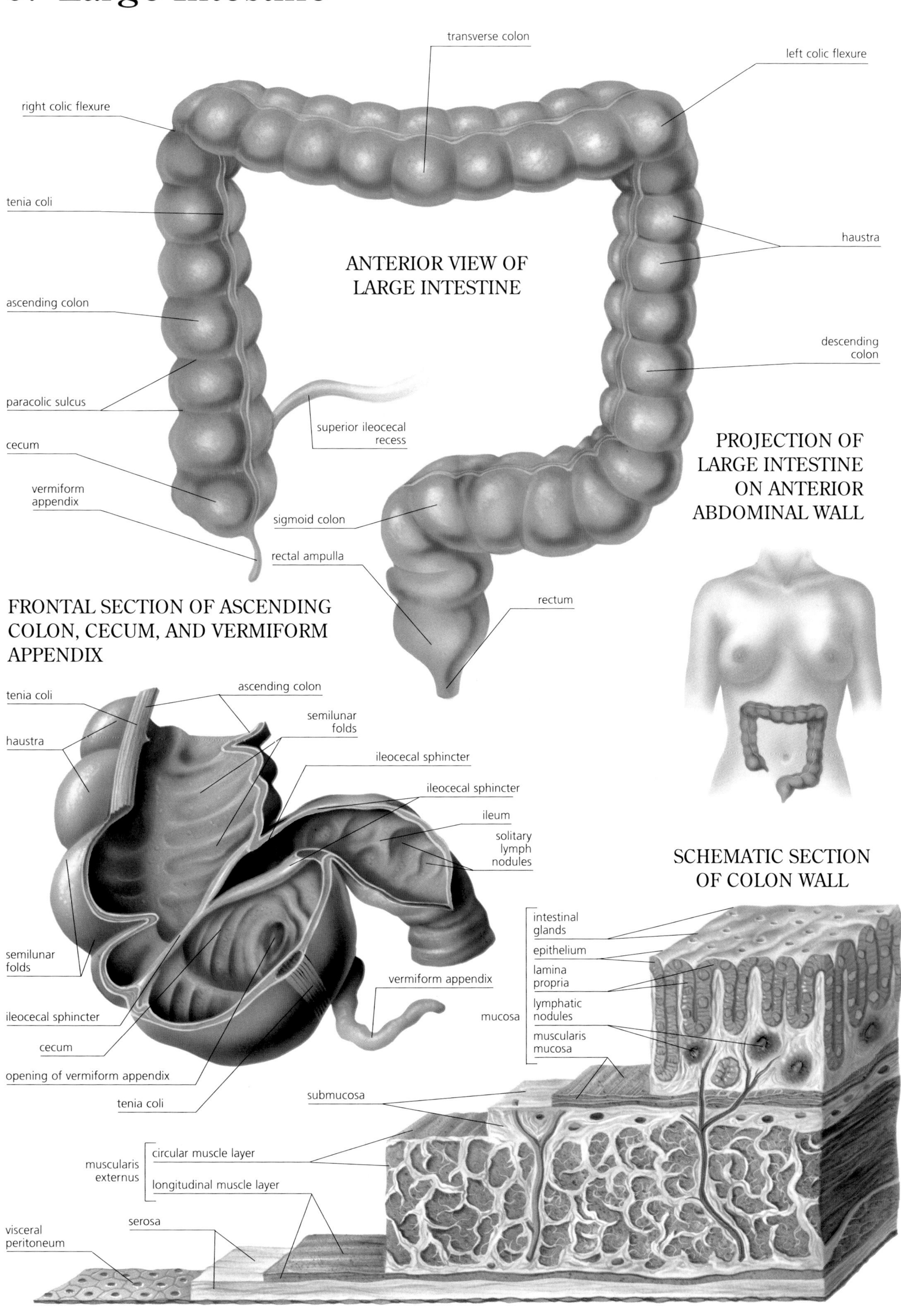

10. Spleen

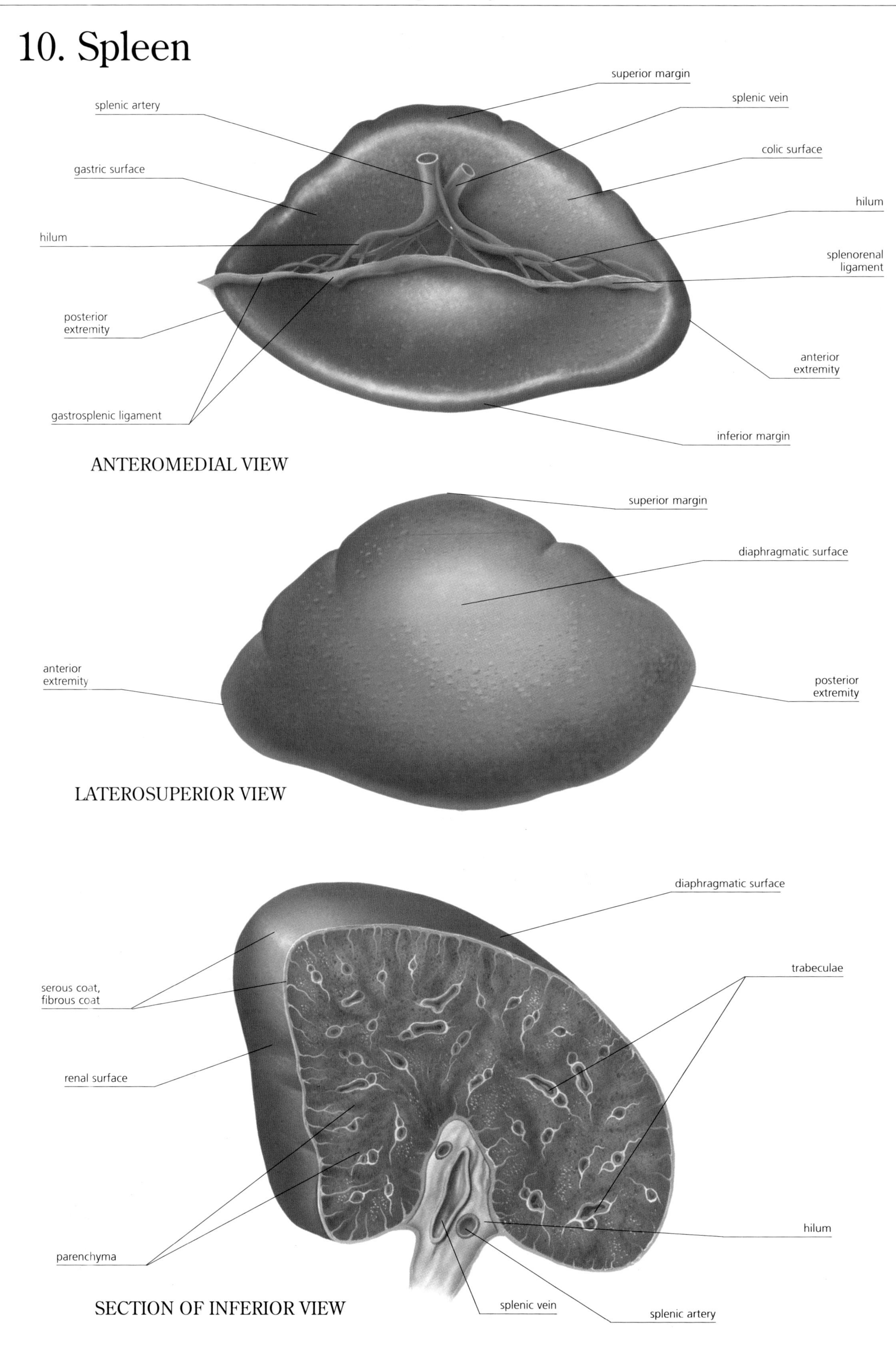

superior margin

splenic artery

splenic vein

gastric surface

colic surface

hilum

hilum

splenorenal ligament

posterior extremity

anterior extremity

gastrosplenic ligament

inferior margin

ANTEROMEDIAL VIEW

superior margin

diaphragmatic surface

anterior extremity

posterior extremity

LATEROSUPERIOR VIEW

diaphragmatic surface

trabeculae

serous coat, fibrous coat

renal surface

hilum

parenchyma

splenic vein

splenic artery

SECTION OF INFERIOR VIEW

11. Teeth

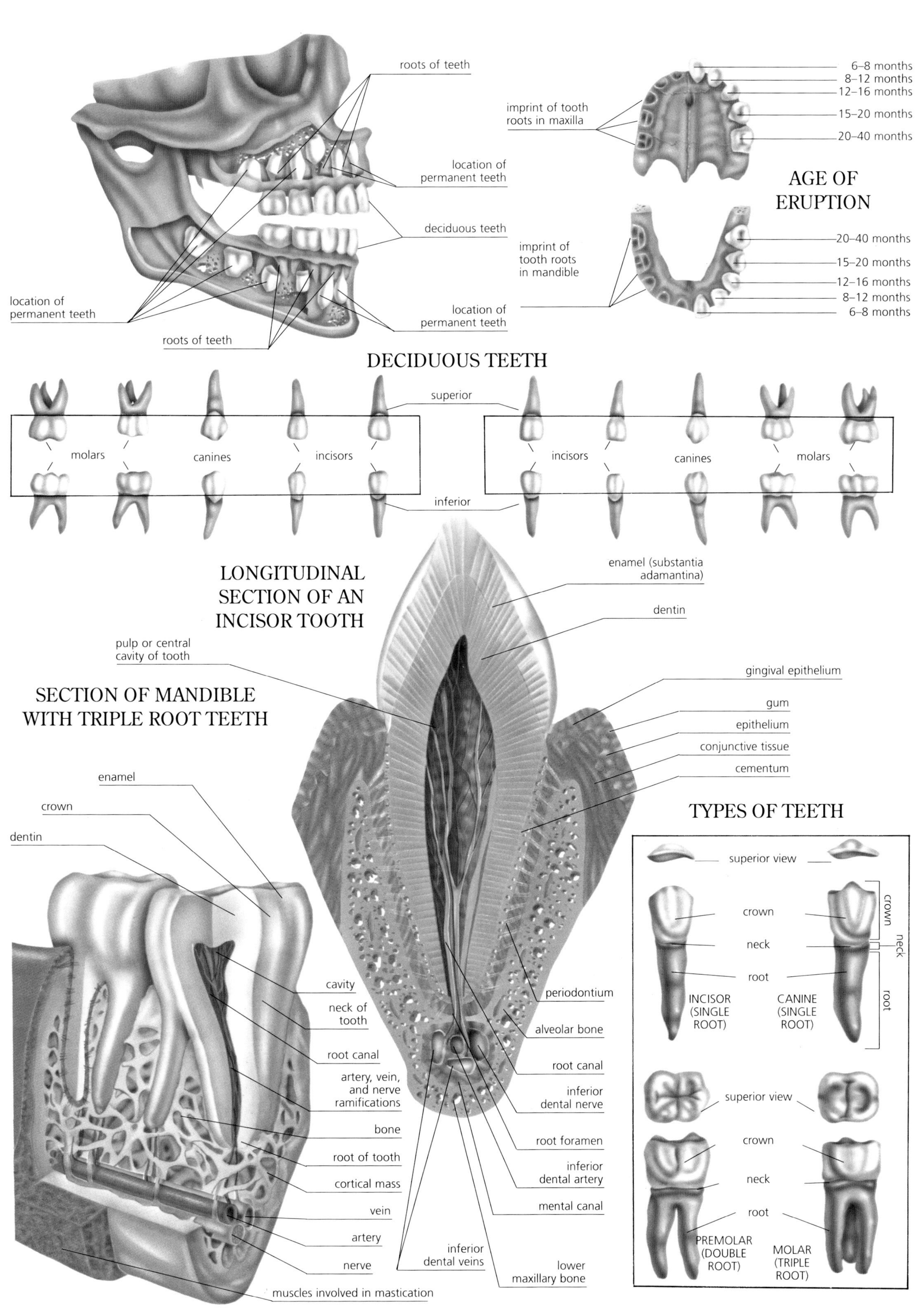

roots of teeth

imprint of tooth roots in maxilla

location of permanent teeth

deciduous teeth

location of permanent teeth

location of permanent teeth

imprint of tooth roots in mandible

roots of teeth

6–8 months
8–12 months
12–16 months
15–20 months
20–40 months

AGE OF ERUPTION

20–40 months
15–20 months
12–16 months
8–12 months
6–8 months

DECIDUOUS TEETH

superior

molars   canines   incisors

inferior

incisors   canines   molars

LONGITUDINAL SECTION OF AN INCISOR TOOTH

pulp or central cavity of tooth

enamel (substantia adamantina)

dentin

gingival epithelium

gum

epithelium

conjunctive tissue

cementum

SECTION OF MANDIBLE WITH TRIPLE ROOT TEETH

enamel

crown

dentin

cavity

neck of tooth

root canal

artery, vein, and nerve ramifications

bone

root of tooth

cortical mass

vein

artery

nerve

muscles involved in mastication

periodontium

alveolar bone

root canal

inferior dental nerve

root foramen

inferior dental artery

mental canal

inferior dental veins

lower maxillary bone

TYPES OF TEETH

superior view

crown

neck

root

INCISOR (SINGLE ROOT)

crown

neck

root

CANINE (SINGLE ROOT)

superior view

crown

neck

root

PREMOLAR (DOUBLE ROOT)

MOLAR (TRIPLE ROOT)

41

12. Teeth

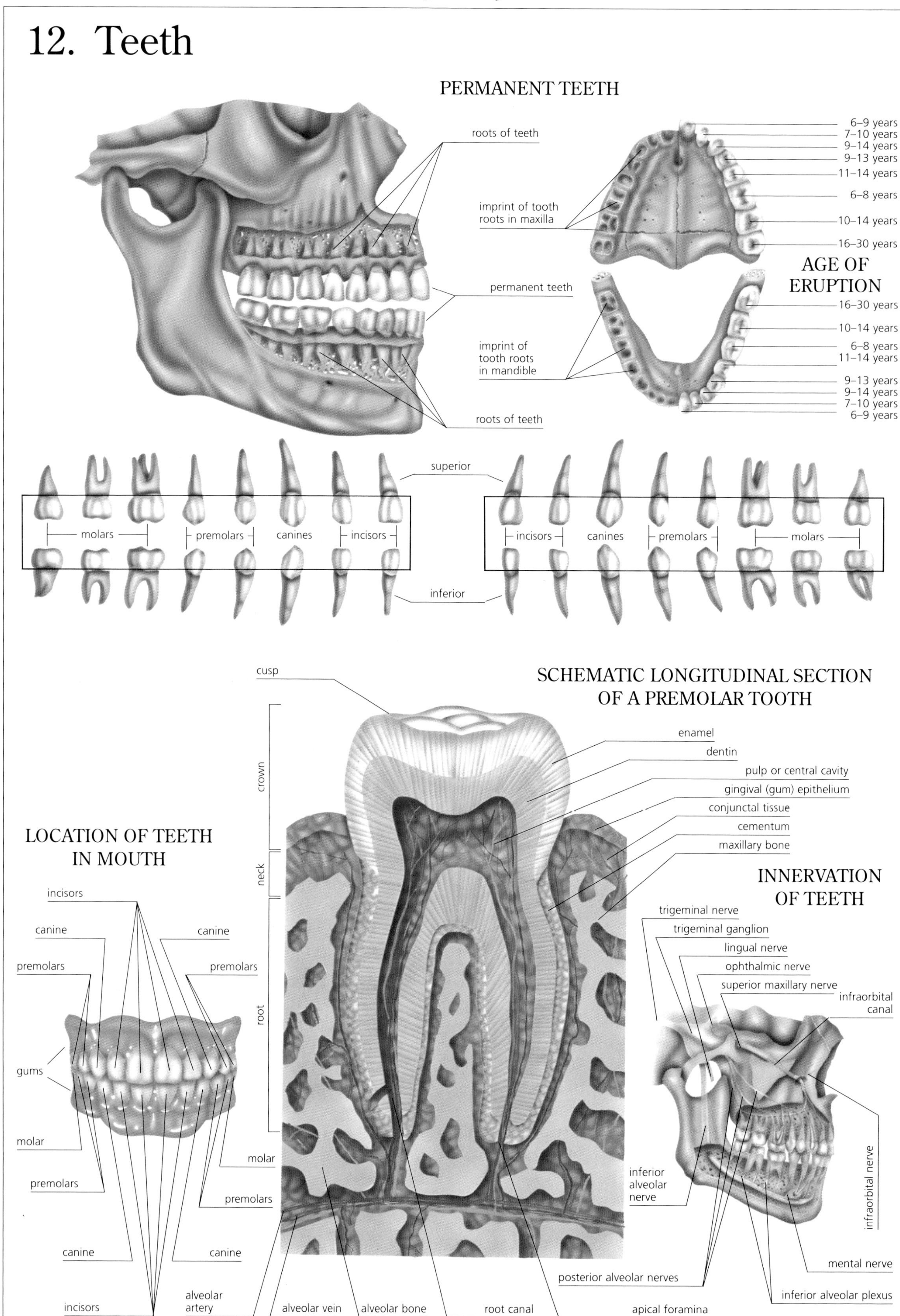

PERMANENT TEETH

roots of teeth

imprint of tooth roots in maxilla

permanent teeth

imprint of tooth roots in mandible

roots of teeth

6–9 years
7–10 years
9–14 years
9–13 years
11–14 years
6–8 years
10–14 years
16–30 years

AGE OF ERUPTION

16–30 years
10–14 years
6–8 years
11–14 years
9–13 years
9–14 years
7–10 years
6–9 years

superior

molars premolars canines incisors

incisors canines premolars molars

inferior

LOCATION OF TEETH IN MOUTH

incisors
canine
premolars
gums
molar
premolars
canine
incisors

canine
premolars
molar
premolars
canine
alveolar artery

cusp

crown
neck
root

alveolar vein alveolar bone root canal

SCHEMATIC LONGITUDINAL SECTION OF A PREMOLAR TOOTH

enamel
dentin
pulp or central cavity
gingival (gum) epithelium
conjunctal tissue
cementum
maxillary bone

INNERVATION OF TEETH

trigeminal nerve
trigeminal ganglion
lingual nerve
ophthalmic nerve
superior maxillary nerve
infraorbital canal

inferior alveolar nerve

posterior alveolar nerves
apical foramina

inferior alveolar plexus
mental nerve

infraorbital nerve

The Respiratory System
1. Lungs and Bronchi

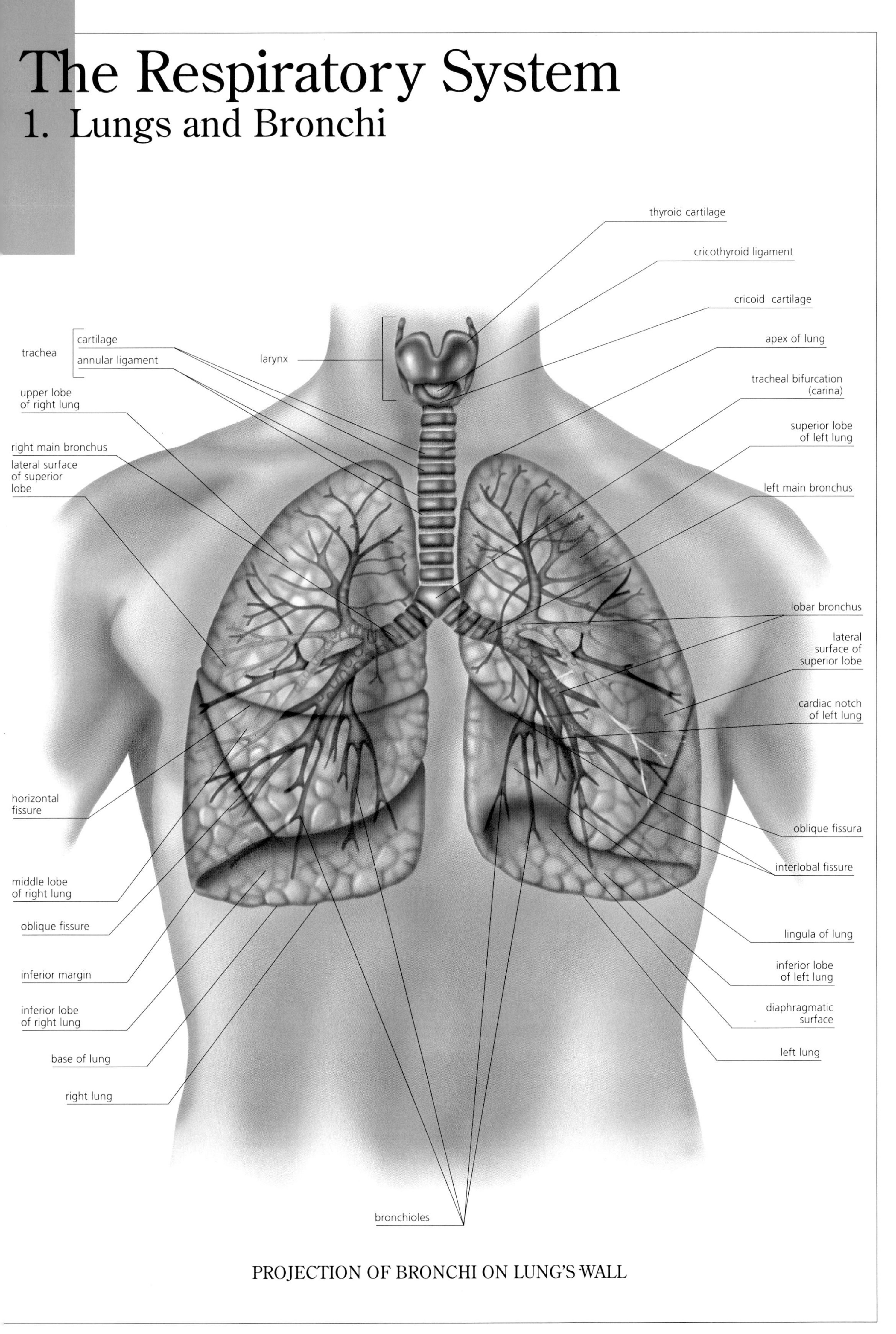

thyroid cartilage

cricothyroid ligament

cricoid cartilage

apex of lung

tracheal bifurcation (carina)

superior lobe of left lung

left main bronchus

lobar bronchus

lateral surface of superior lobe

cardiac notch of left lung

oblique fissura

interlobal fissure

lingula of lung

inferior lobe of left lung

diaphragmatic surface

left lung

trachea

cartilage

annular ligament

larynx

upper lobe of right lung

right main bronchus

lateral surface of superior lobe

horizontal fissure

middle lobe of right lung

oblique fissure

inferior margin

inferior lobe of right lung

base of lung

right lung

bronchioles

PROJECTION OF BRONCHI ON LUNG'S WALL

43

2. Location of Bronchopulmonary Segments

trachea

tracheal bifurcation (carina)

left main bronchus

right main bronchus

lobar bronchus

apical

posterior

anterior

upper lobe of right lung

middle lobe of right lung

lateral

medial

lower lobe of right lung

superior

anterior basal

medial basal

lateral basal

posterior basal

apical

posterior

anterior

superior lingular

inferior lingular

upper lobe of left lung

superior

medial basal

anterior basal

lateral basal

posterior basal

lower lobe of left lung

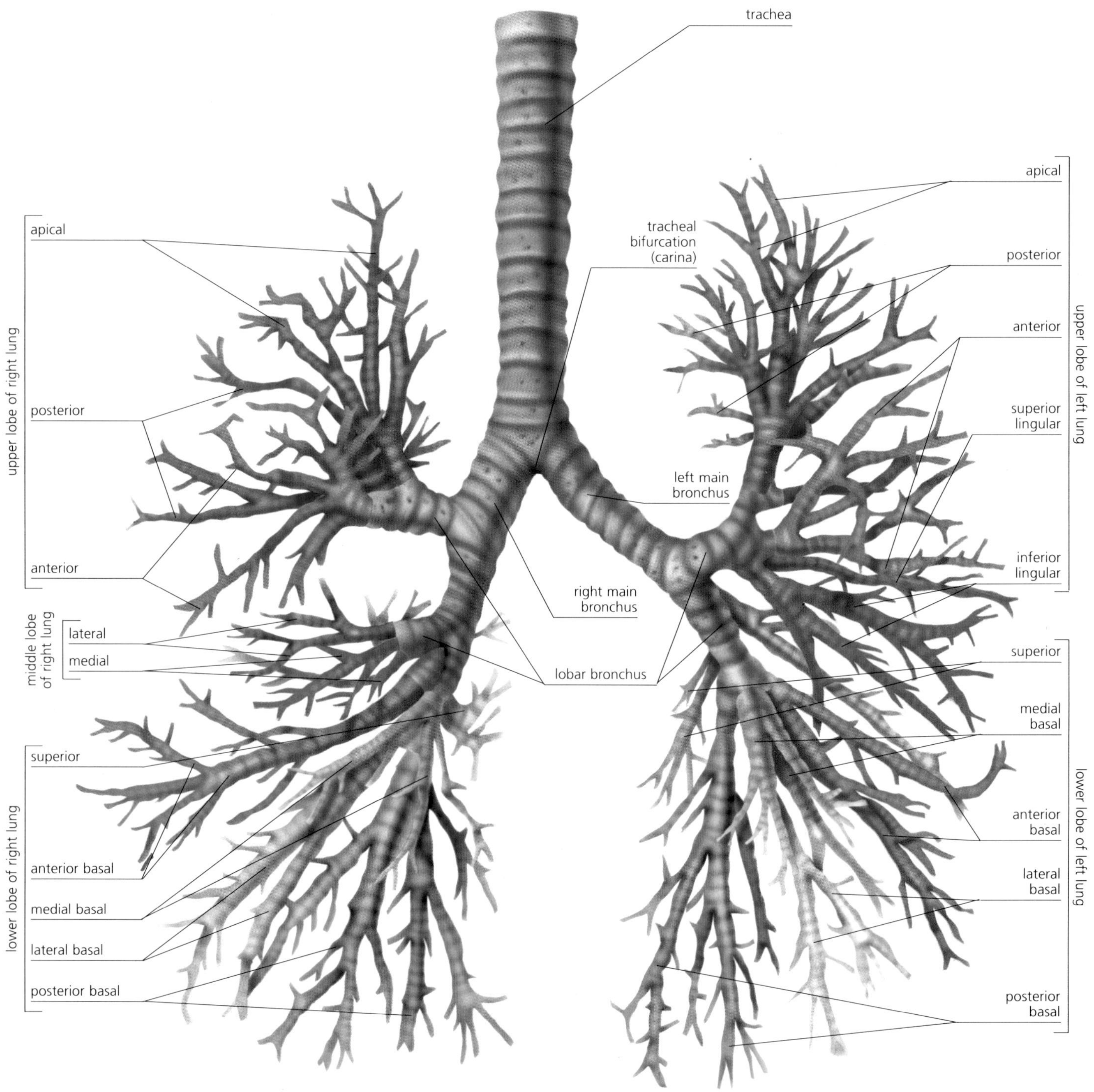

ANTERIOR VIEW OF TRACHEA AND
BRONCHOPULMONARY SEGMENTS
IDENTIFIED BY COLORS

3. Larynx, Trachea, and Bronchi

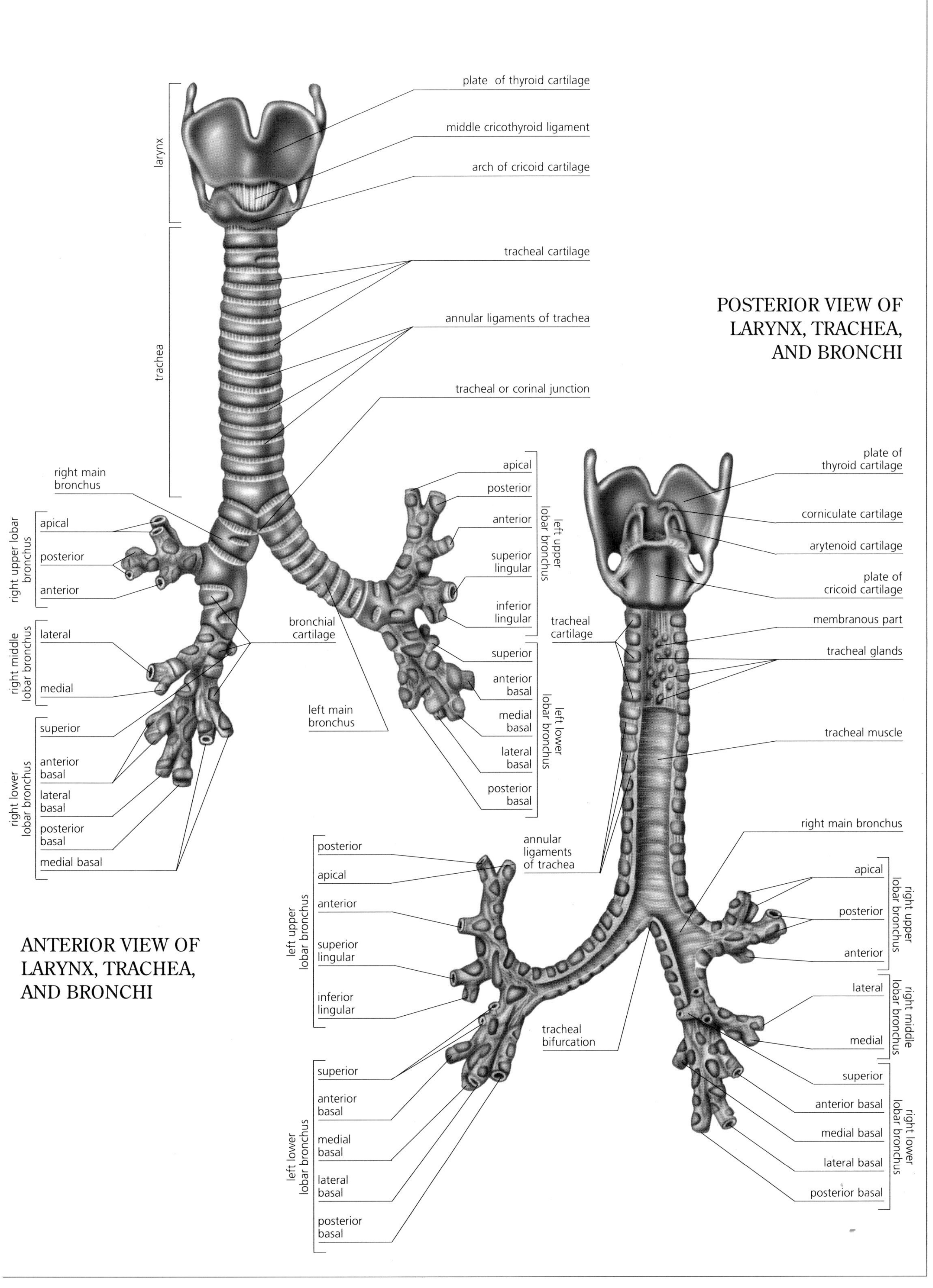

plate of thyroid cartilage

middle cricothyroid ligament

arch of cricoid cartilage

tracheal cartilage

annular ligaments of trachea

tracheal or corinal junction

larynx

trachea

right main bronchus

right upper lobar bronchus

apical

posterior

anterior

right middle lobar bronchus

lateral

medial

right lower lobar bronchus

superior

anterior basal

lateral basal

posterior basal

medial basal

apical

posterior

anterior

superior lingular

inferior lingular

left upper lobar bronchus

superior

anterior basal

medial basal

lateral basal

posterior basal

left lower lobar bronchus

bronchial cartilage

left main bronchus

POSTERIOR VIEW OF LARYNX, TRACHEA, AND BRONCHI

plate of thyroid cartilage

corniculate cartilage

arytenoid cartilage

plate of cricoid cartilage

membranous part

tracheal glands

tracheal muscle

right main bronchus

apical

posterior

anterior

right upper lobar bronchus

lateral

medial

right middle lobar bronchus

superior

anterior basal

medial basal

lateral basal

posterior basal

right lower lobar bronchus

tracheal cartilage

annular ligaments of trachea

tracheal bifurcation

ANTERIOR VIEW OF LARYNX, TRACHEA, AND BRONCHI

posterior

apical

anterior

superior lingular

inferior lingular

left upper lobar bronchus

superior

anterior basal

medial basal

lateral basal

posterior basal

left lower lobar bronchus

4. Breathing. Elements and Mechanism

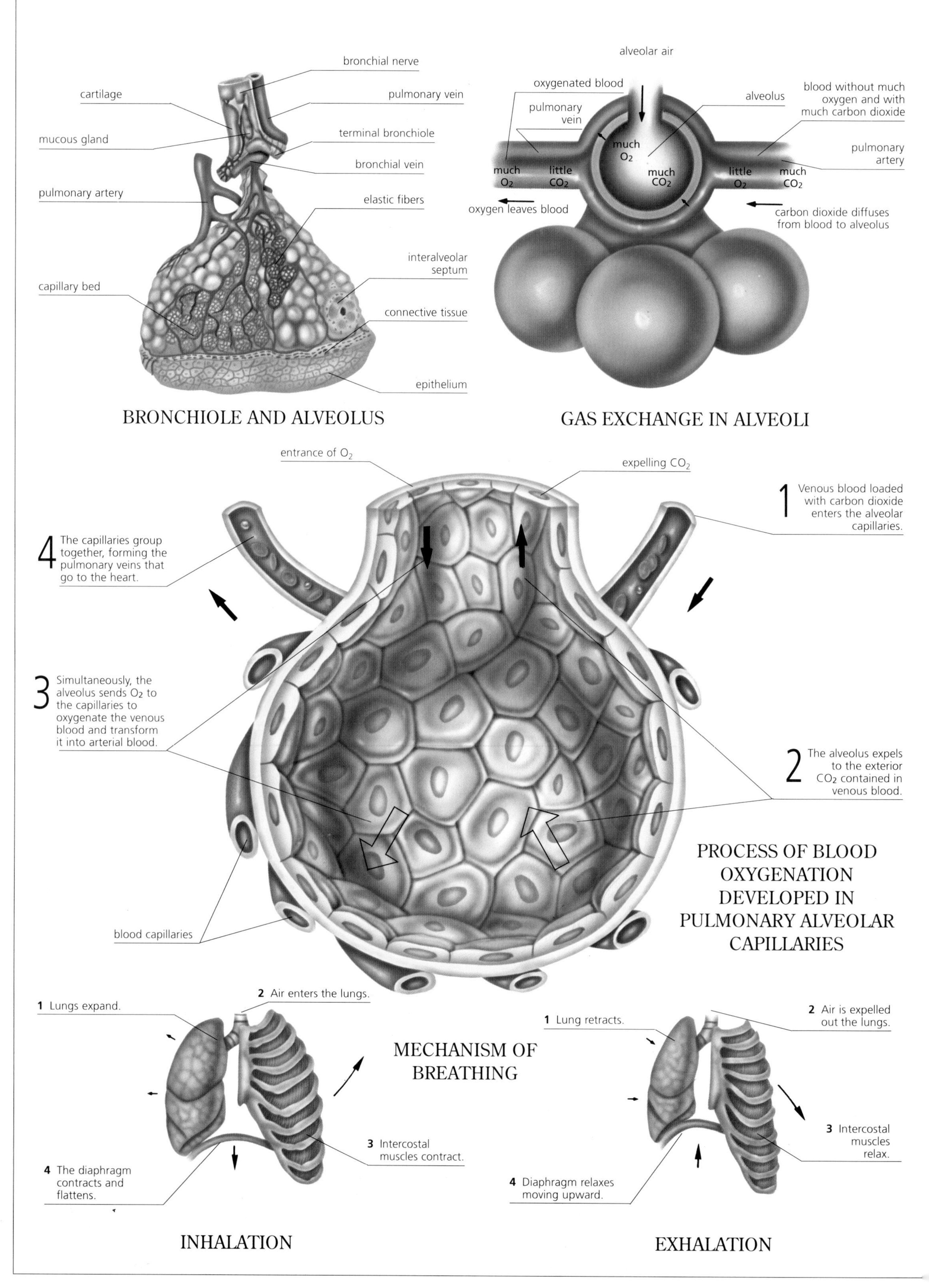

bronchial nerve

cartilage

pulmonary vein

mucous gland

terminal bronchiole

bronchial vein

pulmonary artery

elastic fibers

interalveolar septum

capillary bed

connective tissue

epithelium

BRONCHIOLE AND ALVEOLUS

alveolar air

oxygenated blood

pulmonary vein

alveolus

blood without much oxygen and with much carbon dioxide

much O₂

much O₂

little CO₂

much CO₂

little O₂

much CO₂

pulmonary artery

oxygen leaves blood

carbon dioxide diffuses from blood to alveolus

GAS EXCHANGE IN ALVEOLI

entrance of O₂

expelling CO₂

4 The capillaries group together, forming the pulmonary veins that go to the heart.

1 Venous blood loaded with carbon dioxide enters the alveolar capillaries.

3 Simultaneously, the alveolus sends O₂ to the capillaries to oxygenate the venous blood and transform it into arterial blood.

2 The alveolus expels to the exterior CO₂ contained in venous blood.

blood capillaries

PROCESS OF BLOOD OXYGENATION DEVELOPED IN PULMONARY ALVEOLAR CAPILLARIES

1 Lungs expand.

2 Air enters the lungs.

1 Lung retracts.

2 Air is expelled out the lungs.

MECHANISM OF BREATHING

3 Intercostal muscles contract.

3 Intercostal muscles relax.

4 The diaphragm contracts and flattens.

4 Diaphragm relaxes moving upward.

INHALATION

EXHALATION

The Circulatory System
1. Principal Veins and Arteries

common carotid artery

subclavian artery

arch of aorta

axillary artery

pulmonary artery

coronary artery

brachial artery

hepatic artery

gastric artery

splenic artery

superior mesenteric artery

radial artery

ulnar artery

inferior mesenteric artery

palmar branch

digital artery

bifurcation of aorta

common iliac artery

external iliac artery

internal iliac artery

femoral artery

popliteal artery

fibular artery

anterior tibial artery

posterior tibial artery

dorsal artery of foot

lateral plantar artery

metatarsal artery

internal jugular vein

brachiocephalic vein

subclavian vein

superior vena cava

axillary vein

cephalic vein

pulmonary vein

basilic vein

hepatic vein

intermediate vein

portal hepatic vein

middle ulnar vein

inferior median vein

inferior vena cava

superior mesenteric vein

gastroepiploic vein

inferior mesenteric vein

palmar vein

digital vein

common iliac vein

internal iliac vein

external iliac vein

femoral vein

internal saphenous vein

anterior tibial vein

posterior tibial vein

lateral saphenous vein

dorsal venous arch

digital vein

2. Heart

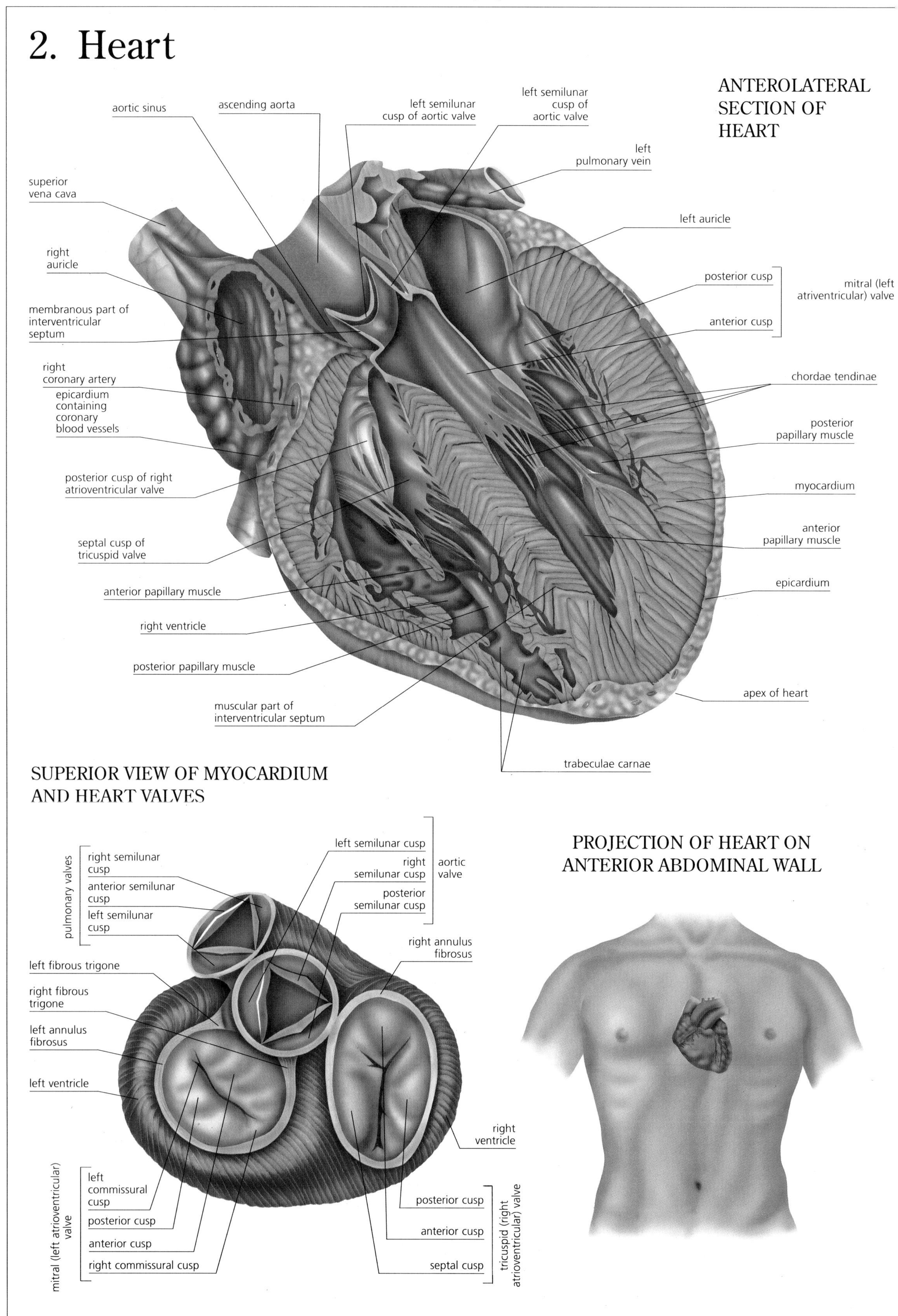

ANTEROLATERAL
SECTION OF
HEART

aortic sinus

ascending aorta

left semilunar
cusp of aortic valve

left semilunar
cusp of
aortic valve

left
pulmonary vein

superior
vena cava

left auricle

right
auricle

posterior cusp

mitral (left
atriventricular) valve

membranous part of
interventricular
septum

anterior cusp

right
coronary artery

chordae tendinae

epicardium
containing
coronary
blood vessels

posterior
papillary muscle

posterior cusp of right
atrioventricular valve

myocardium

septal cusp of
tricuspid valve

anterior
papillary muscle

anterior papillary muscle

epicardium

right ventricle

posterior papillary muscle

apex of heart

muscular part of
interventricular septum

trabeculae carnae

SUPERIOR VIEW OF MYOCARDIUM
AND HEART VALVES

pulmonary valves

right semilunar
cusp

left semilunar cusp

right
semilunar cusp

aortic
valve

anterior semilunar
cusp

posterior
semilunar cusp

left semilunar
cusp

right annulus
fibrosus

left fibrous trigone

right fibrous
trigone

left annulus
fibrosus

left ventricle

right
ventricle

mitral (left atrioventricular) valve

left
commissural
cusp

posterior cusp

anterior cusp

posterior cusp

septal cusp

anterior cusp

right commissural cusp

tricuspid (right atrioventricular) valve

PROJECTION OF HEART ON
ANTERIOR ABDOMINAL WALL

3. Heart

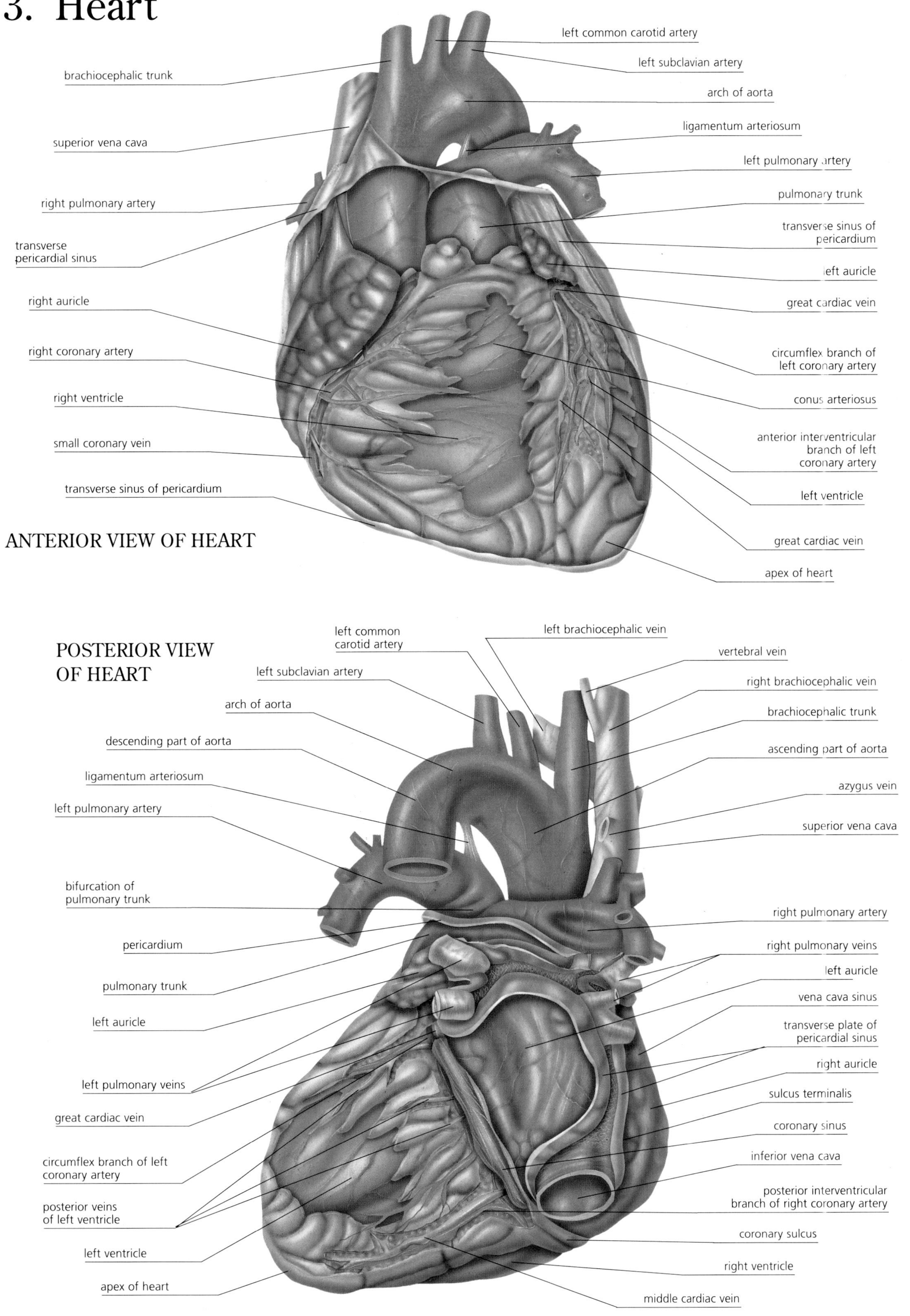

left common carotid artery
left subclavian artery
arch of aorta
ligamentum arteriosum
left pulmonary artery
pulmonary trunk
transverse sinus of pericardium
left auricle
great cardiac vein
circumflex branch of left coronary artery
conus arteriosus
anterior interventricular branch of left coronary artery
left ventricle
great cardiac vein
apex of heart

brachiocephalic trunk
superior vena cava
right pulmonary artery
transverse pericardial sinus
right auricle
right coronary artery
right ventricle
small coronary vein
transverse sinus of pericardium

ANTERIOR VIEW OF HEART

POSTERIOR VIEW OF HEART

left common carotid artery
left subclavian artery
arch of aorta
descending part of aorta
ligamentum arteriosum
left pulmonary artery
bifurcation of pulmonary trunk
pericardium
pulmonary trunk
left auricle
left pulmonary veins
great cardiac vein
circumflex branch of left coronary artery
posterior veins of left ventricle
left ventricle
apex of heart

left brachiocephalic vein
vertebral vein
right brachiocephalic vein
brachiocephalic trunk
ascending part of aorta
azygus vein
superior vena cava
right pulmonary artery
right pulmonary veins
left auricle
vena cava sinus
transverse plate of pericardial sinus
right auricle
sulcus terminalis
coronary sinus
inferior vena cava
posterior interventricular branch of right coronary artery
coronary sulcus
right ventricle
middle cardiac vein

4. Phases of Heartbeat

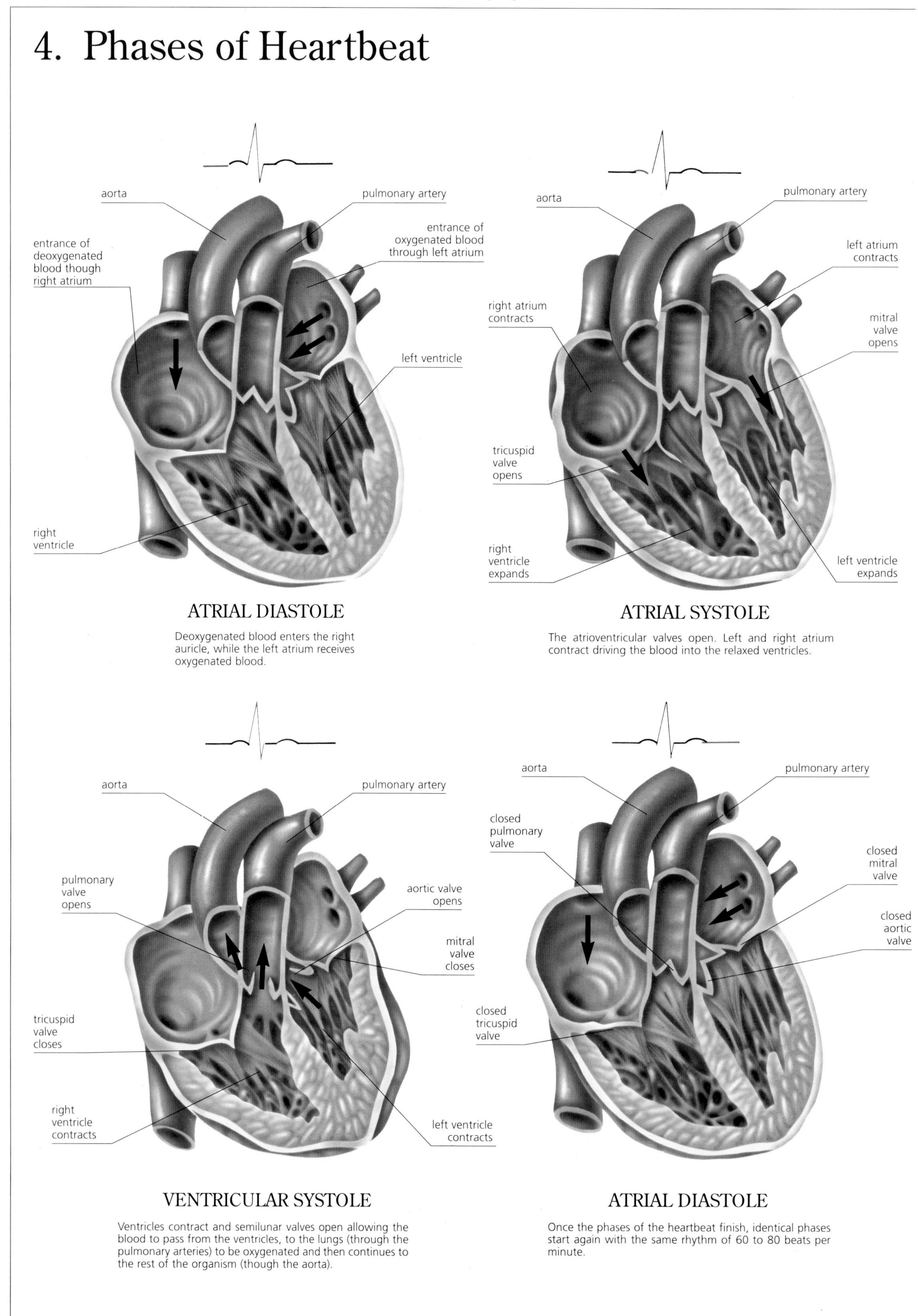

ATRIAL DIASTOLE

Deoxygenated blood enters the right auricle, while the left atrium receives oxygenated blood.

ATRIAL SYSTOLE

The atrioventricular valves open. Left and right atrium contract driving the blood into the relaxed ventricles.

VENTRICULAR SYSTOLE

Ventricles contract and semilunar valves open allowing the blood to pass from the ventricles, to the lungs (through the pulmonary arteries) to be oxygenated and then continues to the rest of the organism (though the aorta).

ATRIAL DIASTOLE

Once the phases of the heartbeat finish, identical phases start again with the same rhythm of 60 to 80 beats per minute.

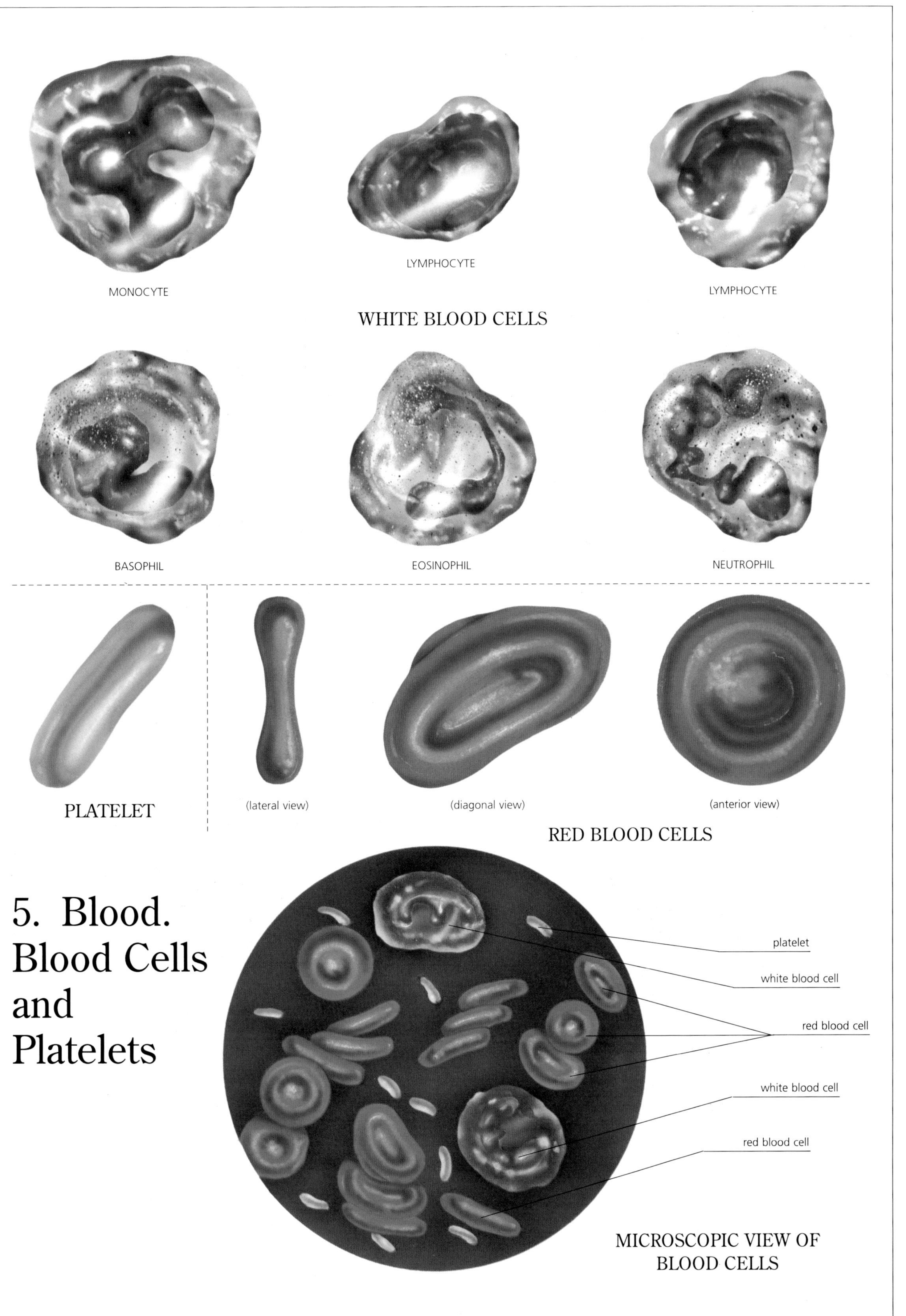

MONOCYTE

LYMPHOCYTE

LYMPHOCYTE

WHITE BLOOD CELLS

BASOPHIL

EOSINOPHIL

NEUTROPHIL

PLATELET

(lateral view)

(diagonal view)

(anterior view)

RED BLOOD CELLS

5. Blood. Blood Cells and Platelets

platelet

white blood cell

red blood cell

white blood cell

red blood cell

MICROSCOPIC VIEW OF
BLOOD CELLS

6. Veins, Arteries, and Valves

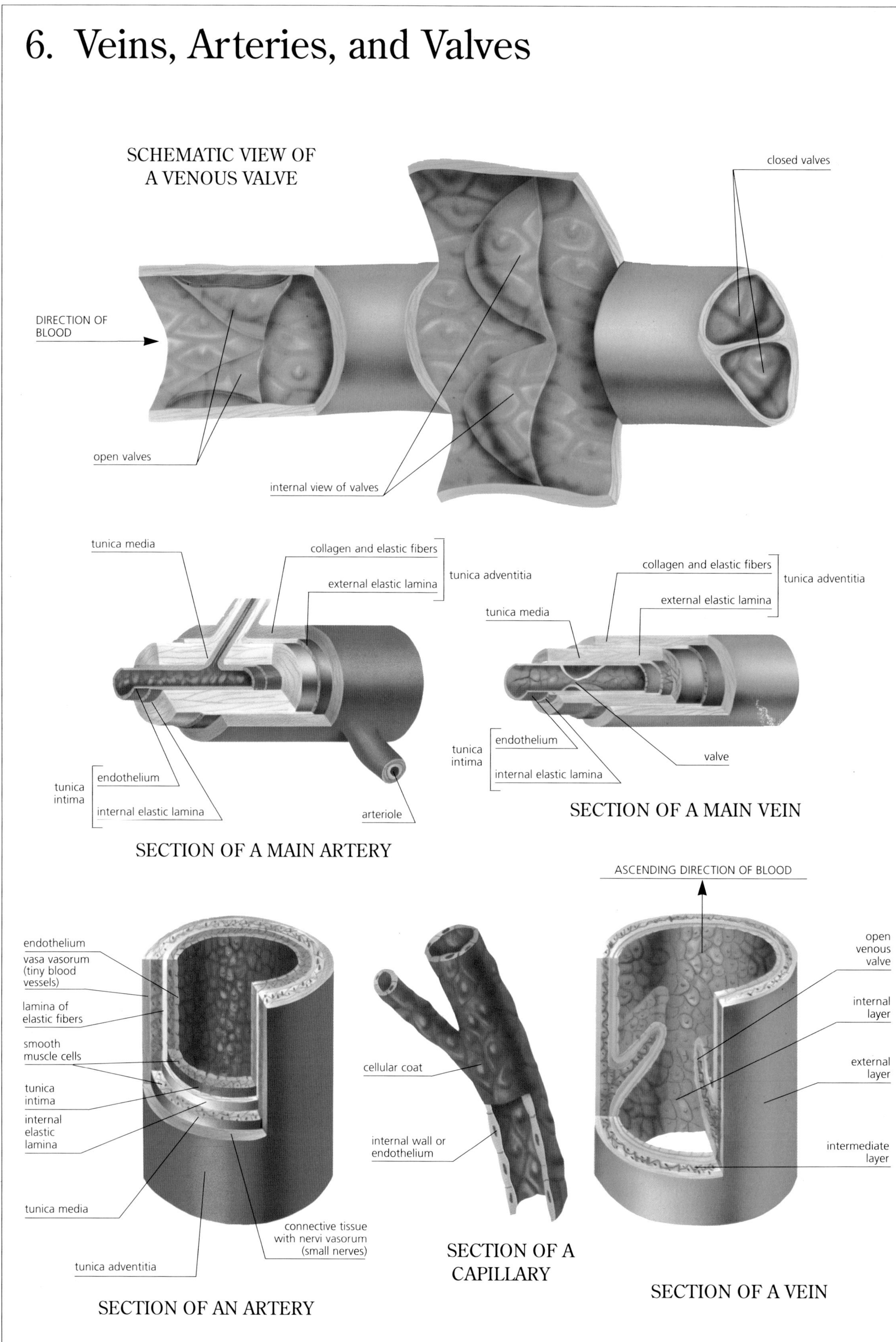

SCHEMATIC VIEW OF
A VENOUS VALVE

closed valves

DIRECTION OF
BLOOD

open valves

internal view of valves

tunica media

collagen and elastic fibers

external elastic lamina

tunica adventitia

endothelium

tunica intima

internal elastic lamina

arteriole

SECTION OF A MAIN ARTERY

collagen and elastic fibers

tunica adventitia

external elastic lamina

tunica media

endothelium

tunica intima

internal elastic lamina

valve

SECTION OF A MAIN VEIN

ASCENDING DIRECTION OF BLOOD

endothelium

vasa vasorum
(tiny blood
vessels)

lamina of
elastic fibers

smooth
muscle cells

tunica
intima

internal
elastic
lamina

tunica media

connective tissue
with nervi vasorum
(small nerves)

tunica adventitia

SECTION OF AN ARTERY

cellular coat

internal wall or
endothelium

SECTION OF A
CAPILLARY

open
venous
valve

internal
layer

external
layer

intermediate
layer

SECTION OF A VEIN

52

The Nervous System
1. Brain. Superior View

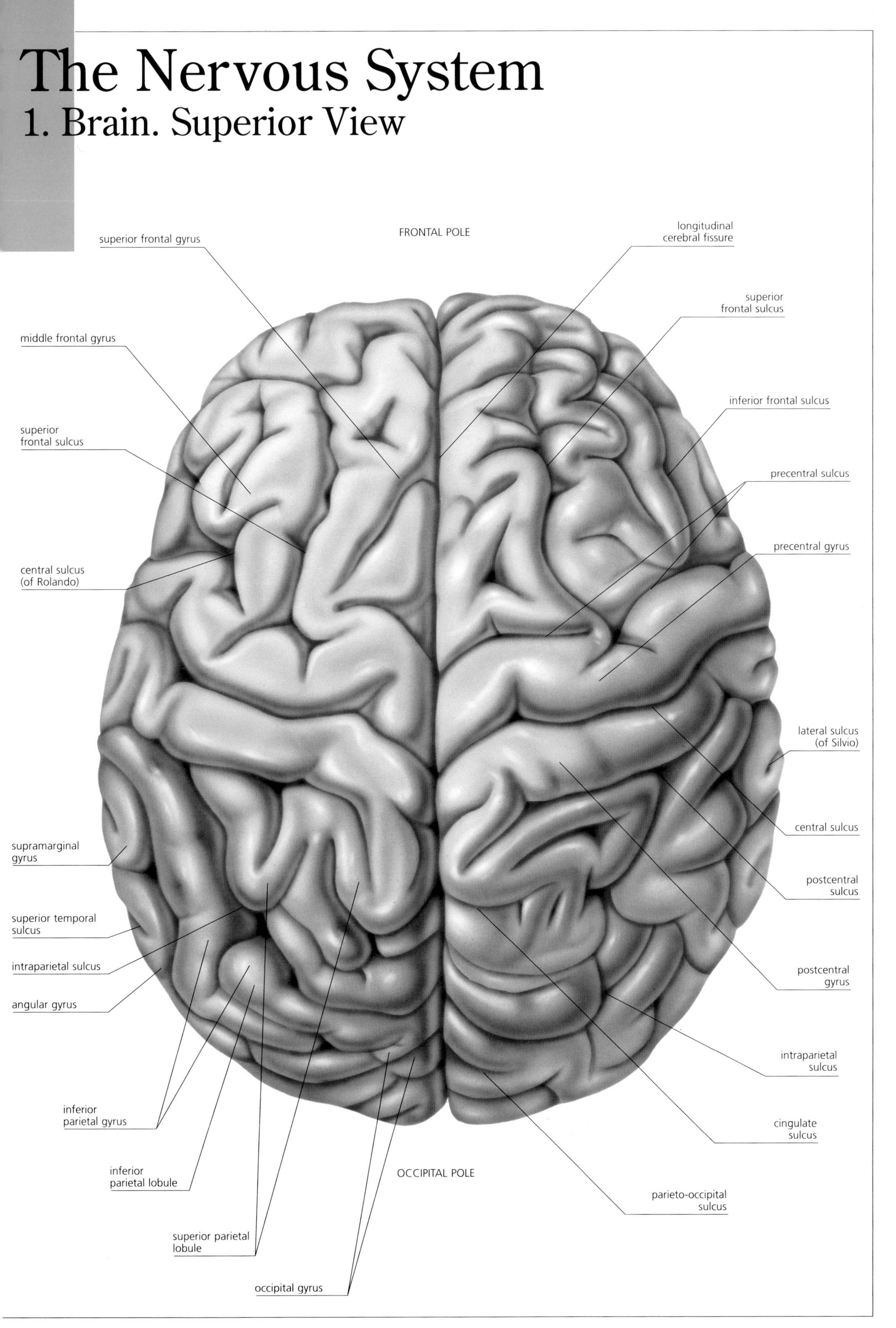

FRONTAL POLE

longitudinal cerebral fissure

superior frontal gyrus

superior frontal sulcus

middle frontal gyrus

inferior frontal sulcus

superior frontal sulcus

precentral sulcus

precentral gyrus

central sulcus (of Rolando)

lateral sulcus (of Silvio)

central sulcus

supramarginal gyrus

postcentral sulcus

superior temporal sulcus

intraparietal sulcus

postcentral gyrus

angular gyrus

intraparietal sulcus

inferior parietal gyrus

cingulate sulcus

inferior parietal lobule

OCCIPITAL POLE

parieto-occipital sulcus

superior parietal lobule

occipital gyrus

2. Brain. Inferior View

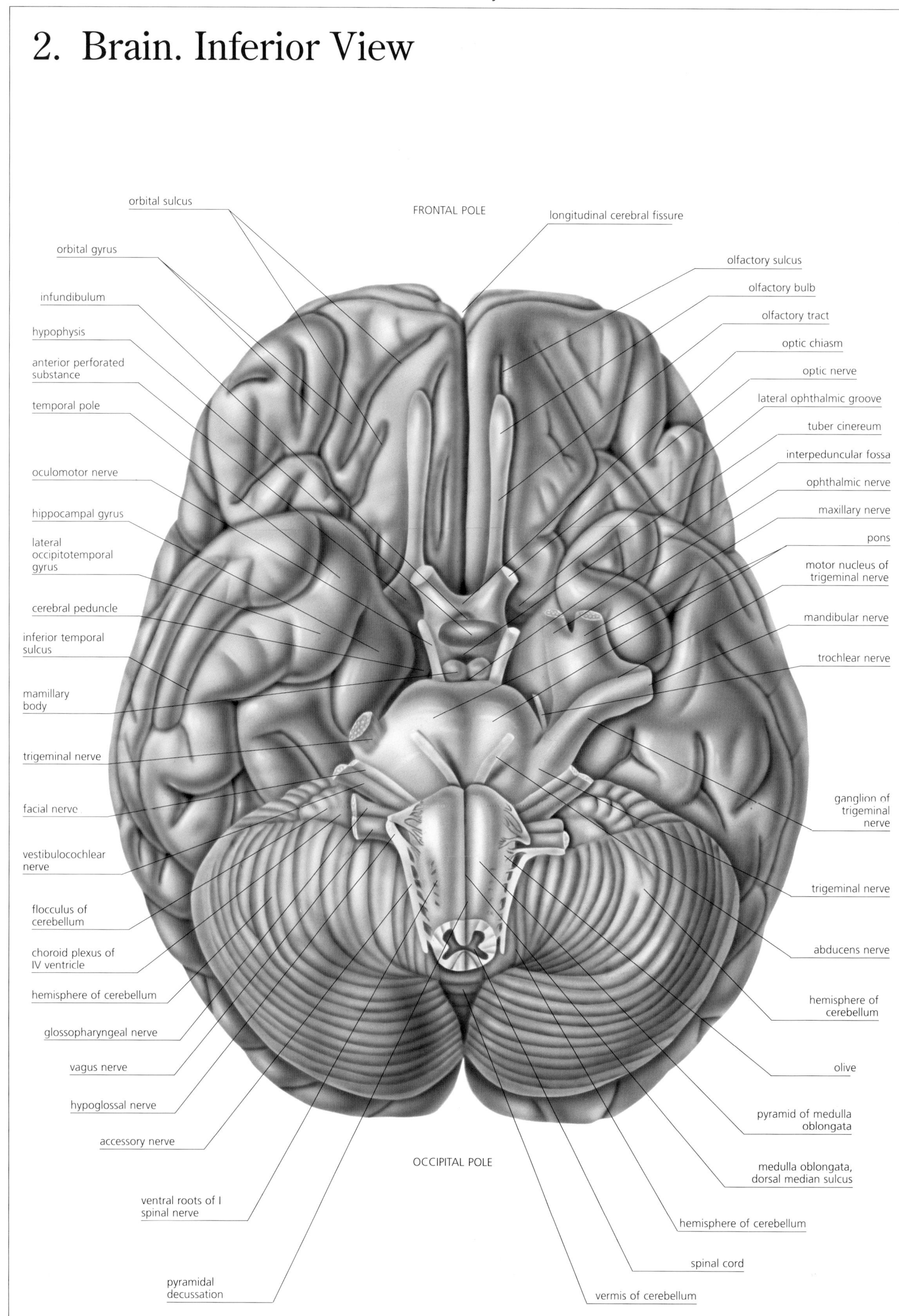

orbital sulcus

FRONTAL POLE

longitudinal cerebral fissure

orbital gyrus

olfactory sulcus

olfactory bulb

infundibulum

olfactory tract

hypophysis

optic chiasm

anterior perforated substance

optic nerve

lateral ophthalmic groove

temporal pole

tuber cinereum

interpeduncular fossa

oculomotor nerve

ophthalmic nerve

maxillary nerve

hippocampal gyrus

pons

lateral occipitotemporal gyrus

motor nucleus of trigeminal nerve

cerebral peduncle

mandibular nerve

inferior temporal sulcus

trochlear nerve

mamillary body

trigeminal nerve

ganglion of trigeminal nerve

facial nerve

vestibulocochlear nerve

trigeminal nerve

flocculus of cerebellum

choroid plexus of IV ventricle

abducens nerve

hemisphere of cerebellum

glossopharyngeal nerve

hemisphere of cerebellum

vagus nerve

olive

hypoglossal nerve

pyramid of medulla oblongata

accessory nerve

OCCIPITAL POLE

medulla oblongata, dorsal median sulcus

ventral roots of I spinal nerve

hemisphere of cerebellum

spinal cord

pyramidal decussation

vermis of cerebellum

3. Peripheral Nervous System

mastoid branch of
superficial cervical plexus

auricular branch of
superficial cervical plexus

transverse cervical nerve

supraclavicular and supracromial
branches of superficial cervical
plexus

dorsal scapular nerve

suprascapular nerve

great serratus nerve

medial pectoral nerve

musculocutaneous nerve

axillary nerve

medial brachial cutaneous nerve, medial
antebrachial cutaneous nerve

median nerve

ulnar nerve

antebrachial
cutaneuous nerve

superficial branch of
radial nerve

dorsal digital branches
of superficial branch of
radial nerve

median nerve

ulnar nerve

sciatic nerve

internal saphenous nerve

posterior tibial nerve

cervical plexus

brachial plexus

posterior fasciculus

lateral fasciculus

middle fasciculus

musculocutaneous nerve

axillary nerve

median nerve

ulnar nerve

radial nerve

iliohypogastric nerve

ilioinguinal nerve

genitofemoral nerve

lateral femoral
cutaneous nerve

femoral nerve

obturator nerve

superior gluteal nerve

lesser sciatic nerve

great sciatic nerve

median nerve

ulnar nerve

sciatic nerve

posterior tibial nerve

pudendal nerve

femoral cutaneous nerve

posterior tibial nerve

superficial peroneal nerve

saphenous nerve

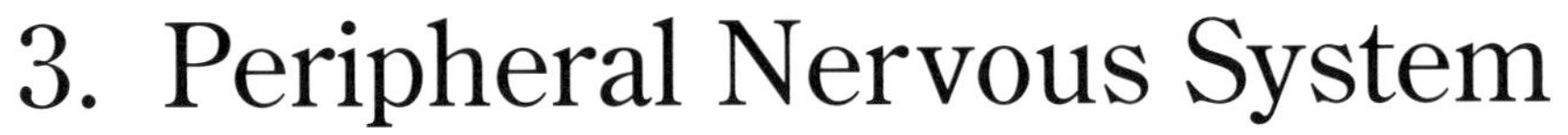

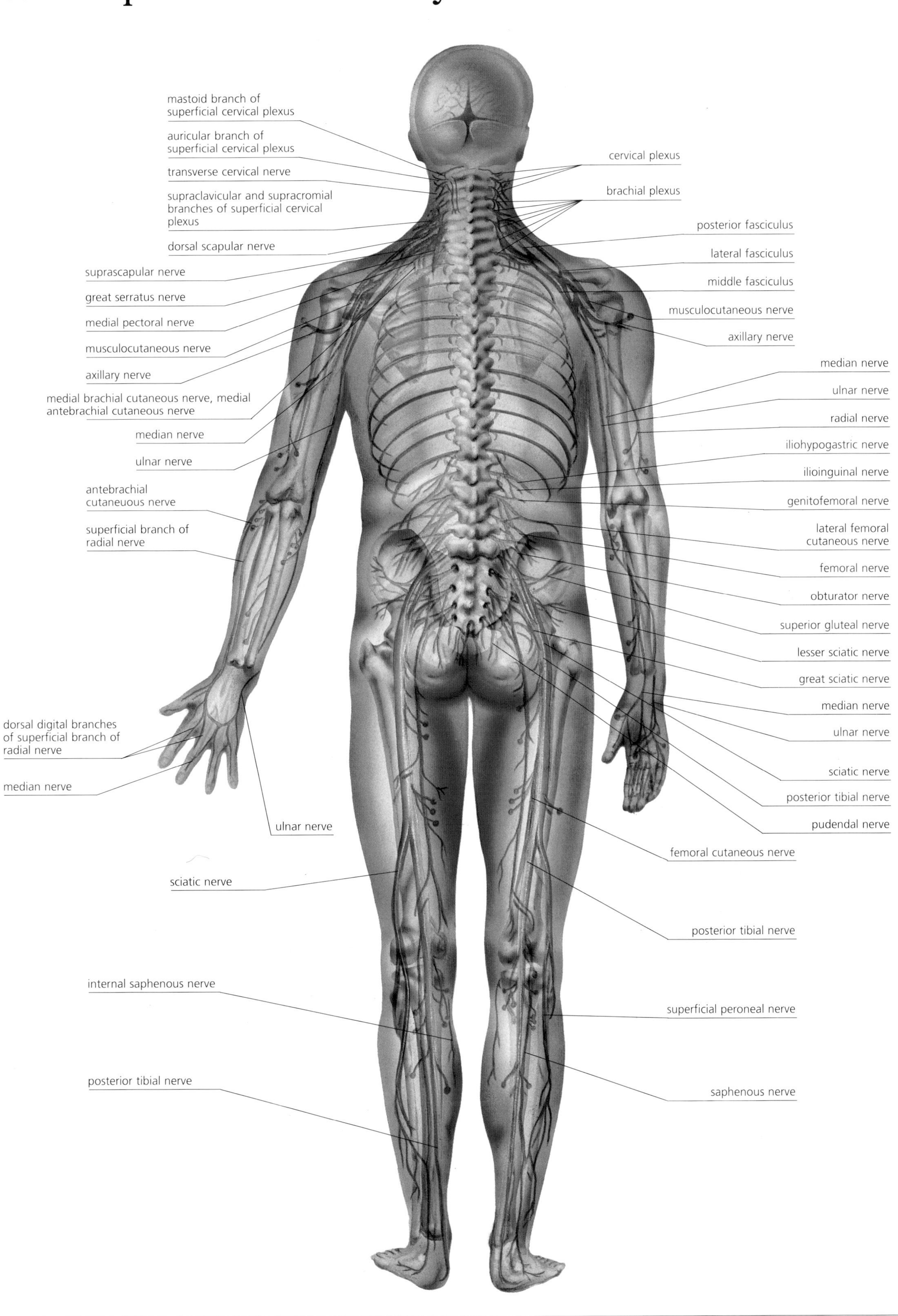

4. Neurons

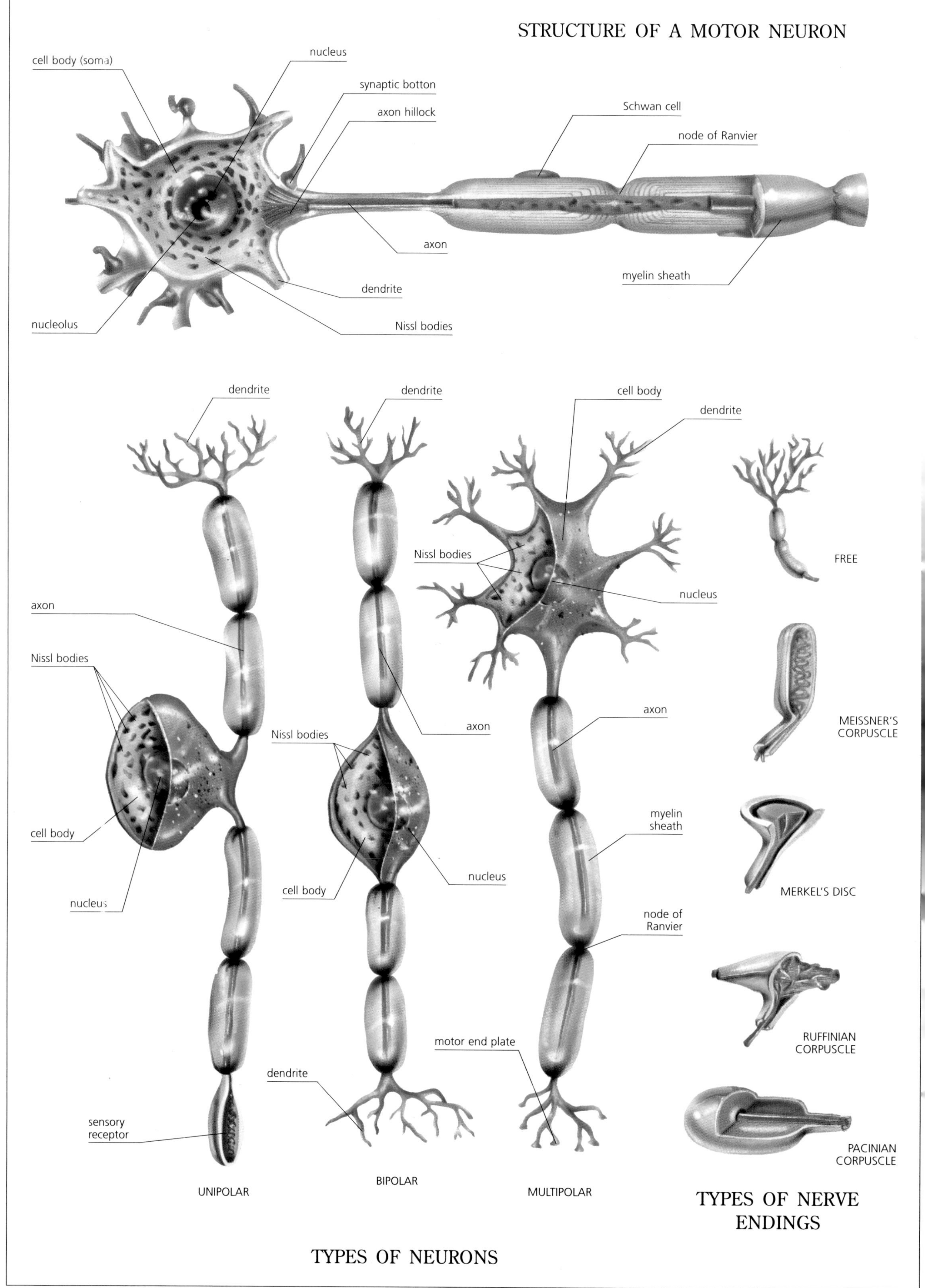

STRUCTURE OF A MOTOR NEURON

cell body (soma)

nucleus

synaptic botton

axon hillock

Schwan cell

node of Ranvier

axon

dendrite

myelin sheath

nucleolus

Nissl bodies

dendrite

dendrite

cell body

dendrite

Nissl bodies

nucleus

FREE

axon

Nissl bodies

axon

MEISSNER'S
CORPUSCLE

cell body

Nissl bodies

axon

myelin
sheath

nucleus

MERKEL'S DISC

cell body

node of
Ranvier

nucleus

RUFFINIAN
CORPUSCLE

motor end plate

sensory
receptor

dendrite

PACINIAN
CORPUSCLE

UNIPOLAR

BIPOLAR

MULTIPOLAR

TYPES OF NERVE
ENDINGS

TYPES OF NEURONS

5. Spinal Cord

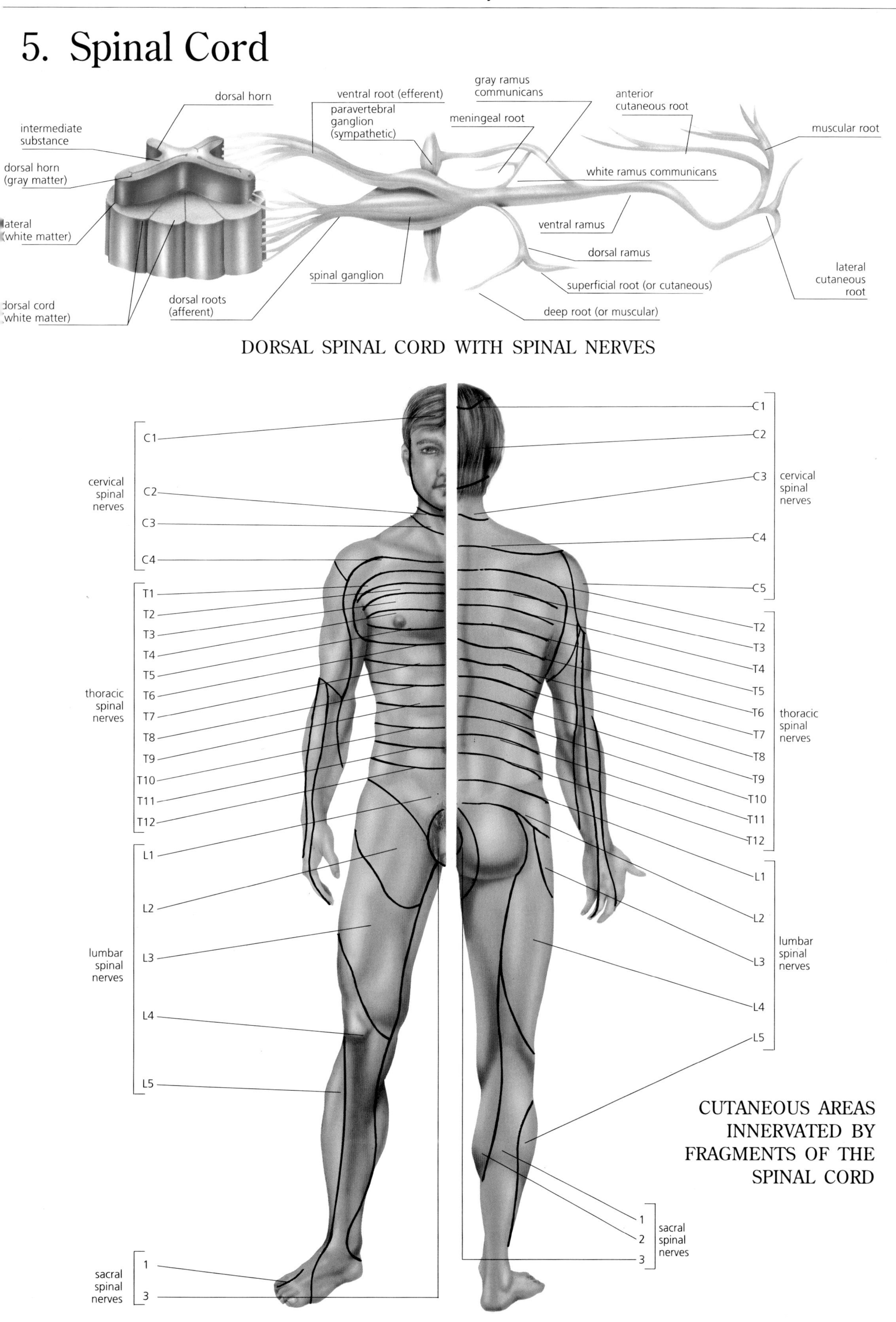

intermediate substance

dorsal horn (gray matter)

lateral (white matter)

dorsal cord (white matter)

dorsal horn

dorsal roots (afferent)

ventral root (efferent)

paravertebral ganglion (sympathetic)

meningeal root

gray ramus communicans

anterior cutaneous root

muscular root

white ramus communicans

ventral ramus

dorsal ramus

spinal ganglion

superficial root (or cutaneous)

deep root (or muscular)

lateral cutaneous root

DORSAL SPINAL CORD WITH SPINAL NERVES

cervical spinal nerves

C1 C2 C3 C4

thoracic spinal nerves

T1 T2 T3 T4 T5 T6 T7 T8 T9 T10 T11 T12

lumbar spinal nerves

L1 L2 L3 L4 L5

sacral spinal nerves

1 3

C1 C2 C3 C4 C5

cervical spinal nerves

T2 T3 T4 T5 T6 T7 T8 T9 T10 T11 T12

thoracic spinal nerves

L1 L2 L3 L4 L5

lumbar spinal nerves

1 2 3 sacral spinal nerves

CUTANEOUS AREAS INNERVATED BY FRAGMENTS OF THE SPINAL CORD

6. Spinal Nerves

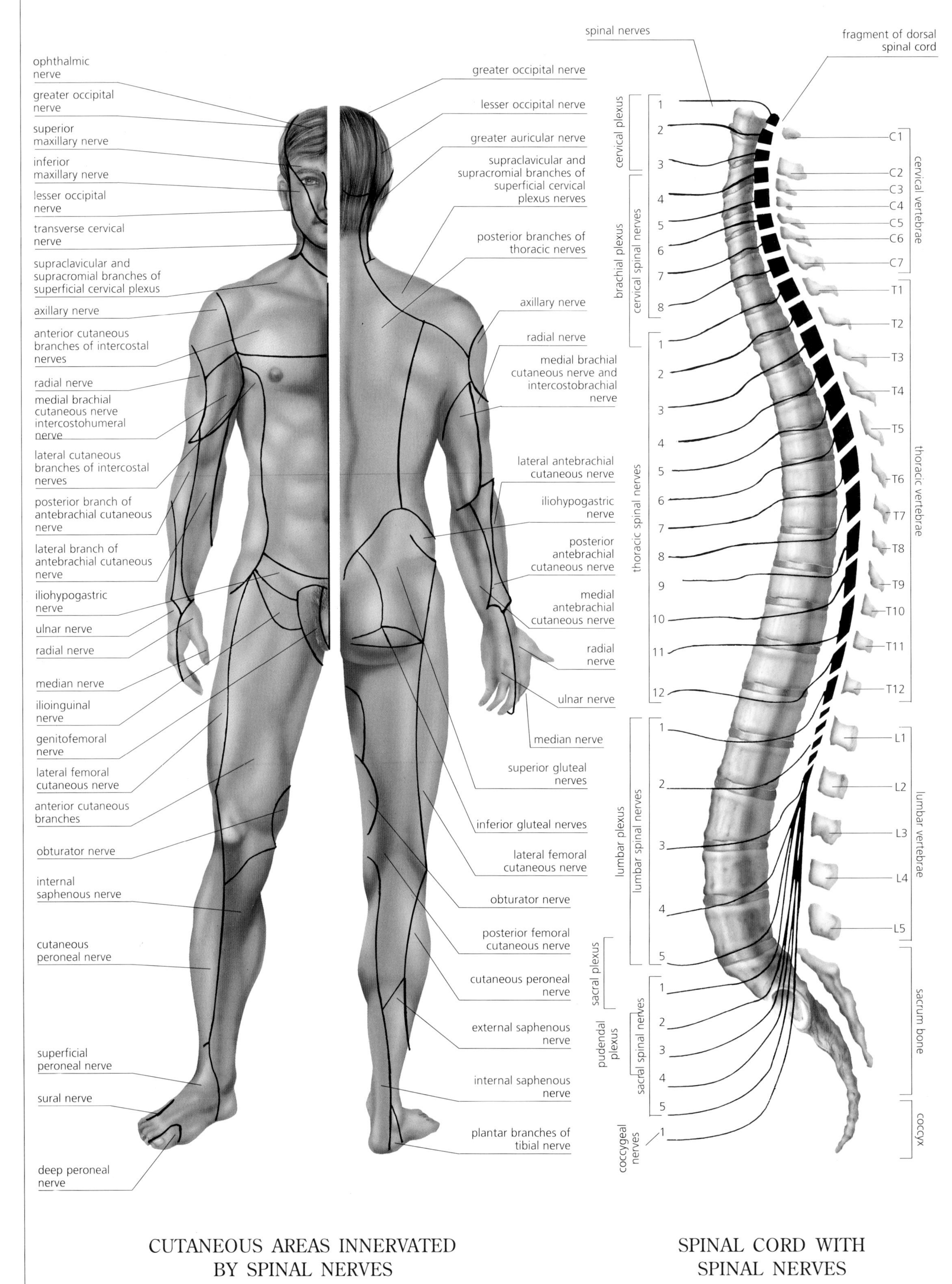

ophthalmic nerve

greater occipital nerve

superior maxillary nerve

inferior maxillary nerve

lesser occipital nerve

transverse cervical nerve

supraclavicular and supracromial branches of superficial cervical plexus

axillary nerve

anterior cutaneous branches of intercostal nerves

radial nerve

medial brachial cutaneous nerve intercostohumeral nerve

lateral cutaneous branches of intercostal nerves

posterior branch of antebrachial cutaneous nerve

lateral branch of antebrachial cutaneous nerve

iliohypogastric nerve

ulnar nerve

radial nerve

median nerve

ilioinguinal nerve

genitofemoral nerve

lateral femoral cutaneous nerve

anterior cutaneous branches

obturator nerve

internal saphenous nerve

cutaneous peroneal nerve

superficial peroneal nerve

sural nerve

deep peroneal nerve

greater occipital nerve

lesser occipital nerve

greater auricular nerve

supraclavicular and supracromial branches of superficial cervical plexus nerves

posterior branches of thoracic nerves

axillary nerve

radial nerve

medial brachial cutaneous nerve and intercostobrachial nerve

lateral antebrachial cutaneous nerve

iliohypogastric nerve

posterior antebrachial cutaneous nerve

medial antebrachial cutaneous nerve

radial nerve

ulnar nerve

median nerve

superior gluteal nerves

inferior gluteal nerves

lateral femoral cutaneous nerve

obturator nerve

posterior femoral cutaneous nerve

cutaneous peroneal nerve

external saphenous nerve

internal saphenous nerve

plantar branches of tibial nerve

spinal nerves

fragment of dorsal spinal cord

cervical plexus

brachial plexus

cervical spinal nerves

1 2 3 4 5 6 7 8

thoracic spinal nerves

1 2 3 4 5 6 7 8 9 10 11 12

lumbar plexus

lumbar spinal nerves

1 2 3 4 5

sacral plexus

pudendal plexus

sacral spinal nerves

1 2 3 4 5

coccygeal nerves

1

C1 C2 C3 C4 C5 C6 C7

cervical vertebrae

T1 T2 T3 T4 T5 T6 T7 T8 T9 T10 T11 T12

thoracic vertebrae

L1 L2 L3 L4 L5

lumbar vertebrae

sacrum bone

coccyx

CUTANEOUS AREAS INNERVATED
BY SPINAL NERVES

SPINAL CORD WITH
SPINAL NERVES

The Senses
1. Sight

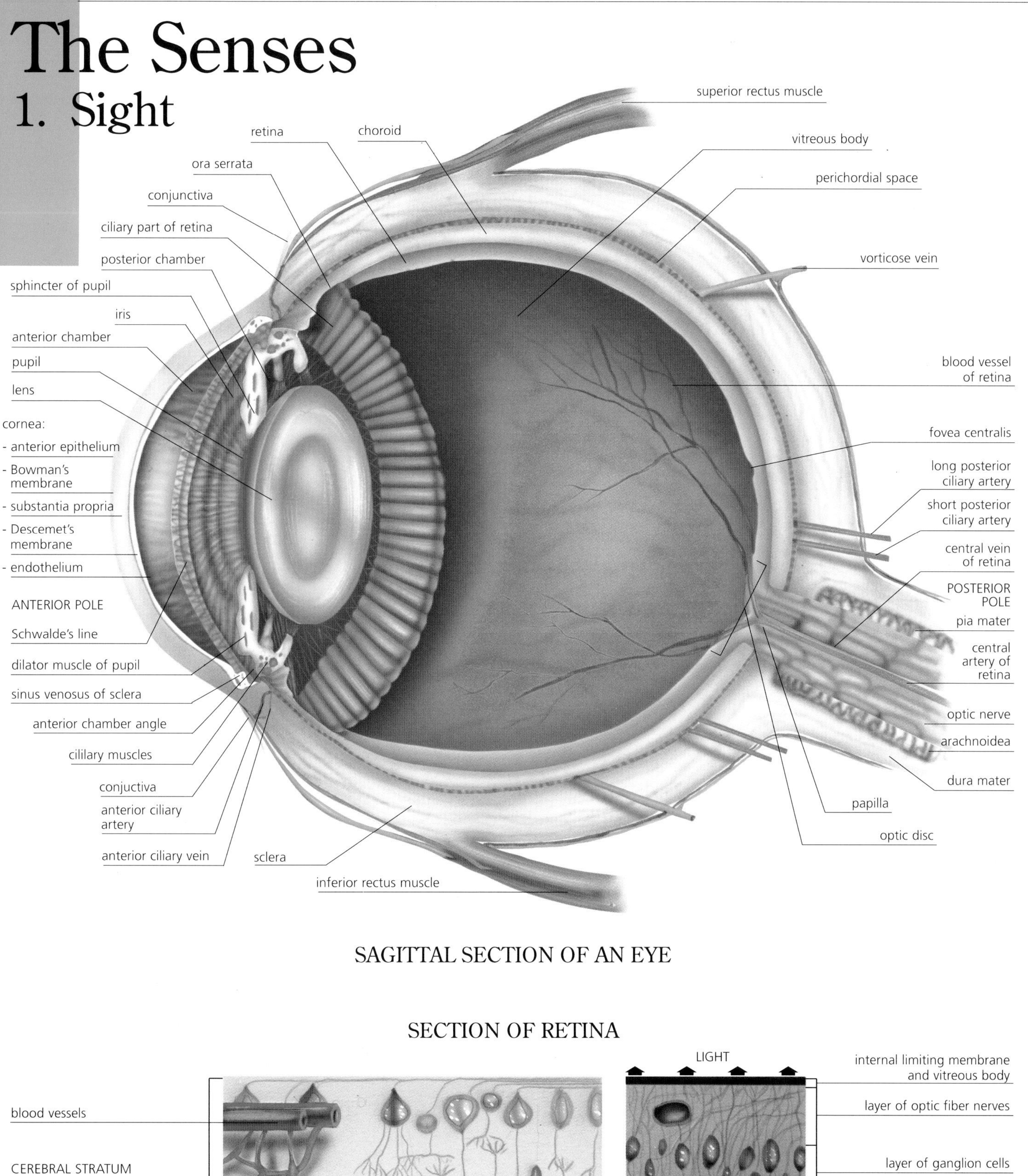

superior rectus muscle

vitreous body

perichordial space

vorticose vein

retina
choroid

ora serrata

conjunctiva

ciliary part of retina

posterior chamber

sphincter of pupil

iris

anterior chamber

pupil

lens

cornea:
- anterior epithelium
- Bowman's membrane
- substantia propria
- Descemet's membrane
- endothelium

ANTERIOR POLE

Schwalde's line

dilator muscle of pupil

sinus venosus of sclera

anterior chamber angle

cililary muscles

conjuctiva

anterior ciliary artery

anterior ciliary vein

sclera

inferior rectus muscle

blood vessel of retina

fovea centralis

long posterior ciliary artery

short posterior ciliary artery

central vein of retina

POSTERIOR POLE

pia mater

central artery of retina

optic nerve

arachnoidea

dura mater

papilla

optic disc

SAGITTAL SECTION OF AN EYE

SECTION OF RETINA

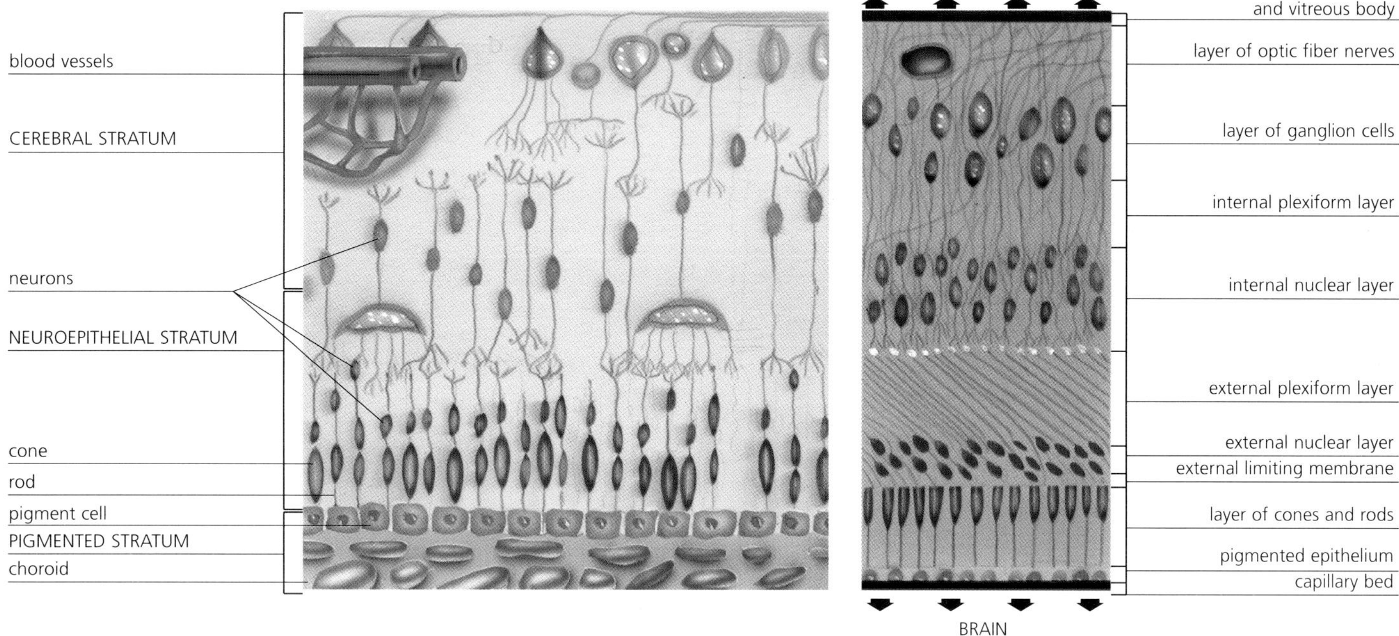

LIGHT

blood vessels

CEREBRAL STRATUM

neurons

NEUROEPITHELIAL STRATUM

cone

rod

pigment cell

PIGMENTED STRATUM

choroid

internal limiting membrane and vitreous body

layer of optic fiber nerves

layer of ganglion cells

internal plexiform layer

internal nuclear layer

external plexiform layer

external nuclear layer

external limiting membrane

layer of cones and rods

pigmented epithelium

capillary bed

BRAIN

2. Sight

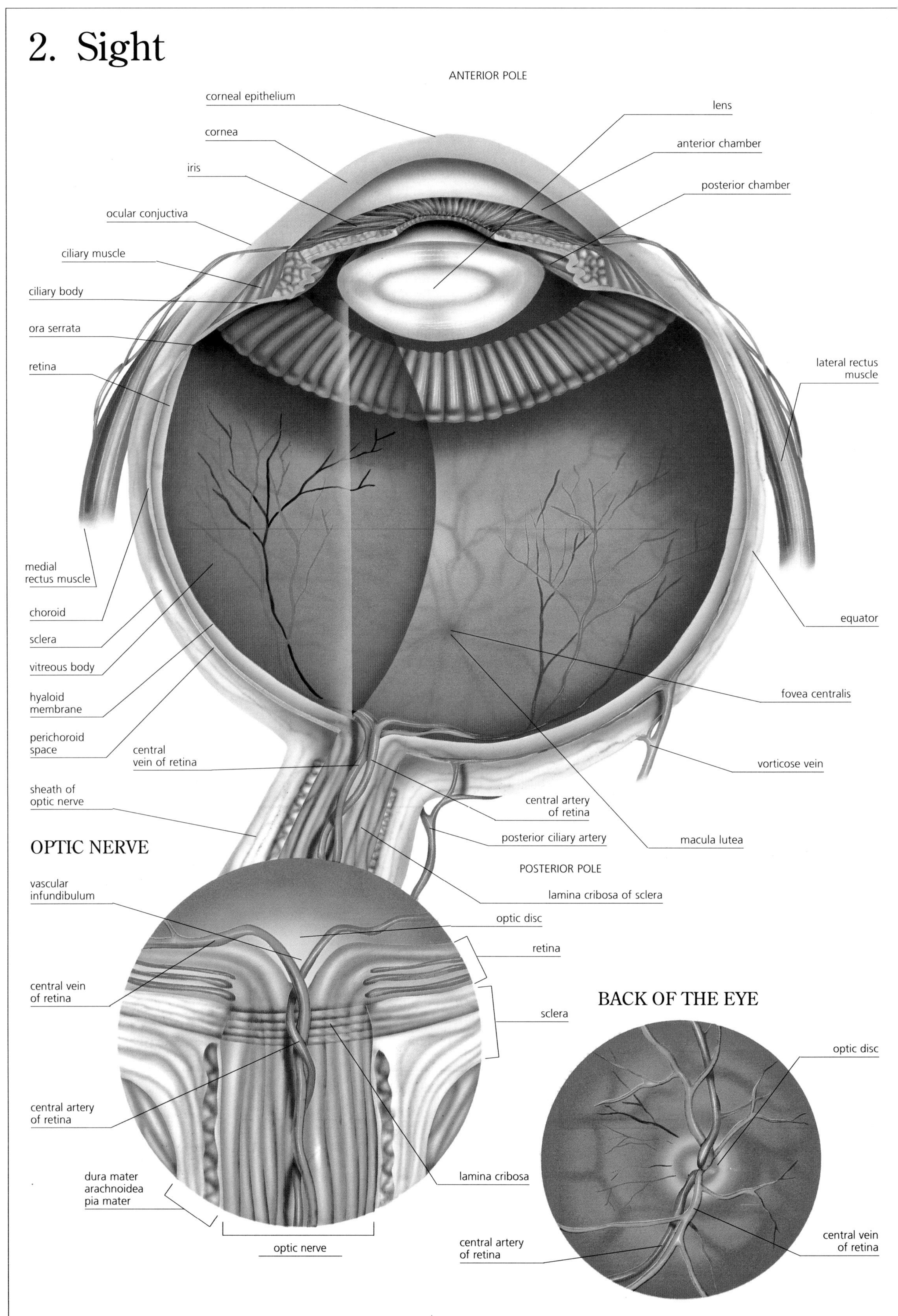

ANTERIOR POLE

corneal epithelium

cornea

iris

ocular conjuctiva

ciliary muscle

ciliary body

ora serrata

retina

medial rectus muscle

choroid

sclera

vitreous body

hyaloid membrane

perichoroid space

central vein of retina

sheath of optic nerve

OPTIC NERVE

vascular infundibulum

central vein of retina

central artery of retina

dura mater
arachnoidea
pia mater

optic nerve

lens

anterior chamber

posterior chamber

lateral rectus muscle

equator

fovea centralis

vorticose vein

central artery of retina

posterior ciliary artery

macula lutea

POSTERIOR POLE

lamina cribosa of sclera

optic disc

retina

sclera

BACK OF THE EYE

lamina cribosa

central artery of retina

optic disc

central vein of retina

3. Sight

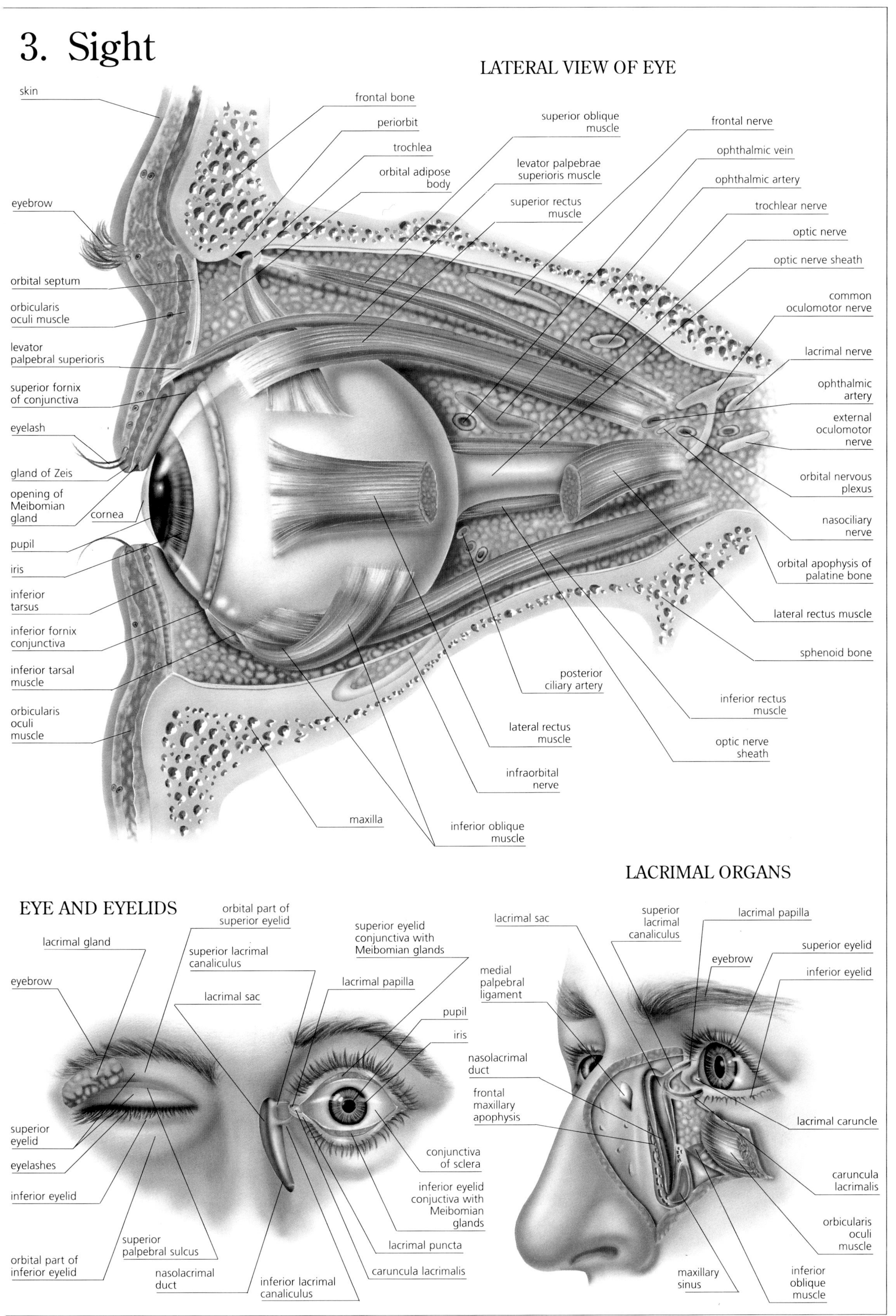

LATERAL VIEW OF EYE

skin
eyebrow
orbital septum
orbicularis
oculi muscle
levator
palpebral superioris
superior fornix
of conjunctiva
eyelash
gland of Zeis
opening of
Meibomian
gland
cornea
pupil
iris
inferior
tarsus
inferior fornix
conjunctiva
inferior tarsal
muscle
orbicularis
oculi
muscle

frontal bone
periorbit
trochlea
orbital adipose
body

superior oblique
muscle
levator palpebrae
superioris muscle
superior rectus
muscle

frontal nerve
ophthalmic vein
ophthalmic artery
trochlear nerve
optic nerve
optic nerve sheath
common
oculomotor nerve
lacrimal nerve
ophthalmic
artery
external
oculomotor
nerve
orbital nervous
plexus
nasociliary
nerve
orbital apophysis of
palatine bone
lateral rectus muscle
sphenoid bone
inferior rectus
muscle
optic nerve
sheath

posterior
ciliary artery
lateral rectus
muscle
infraorbital
nerve
maxilla
inferior oblique
muscle

LACRIMAL ORGANS

EYE AND EYELIDS

lacrimal gland
eyebrow
superior
eyelid
eyelashes
inferior eyelid
orbital part of
inferior eyelid
nasolacrimal
duct

orbital part of
superior eyelid
superior lacrimal
canaliculus
lacrimal sac
superior palpebral
sulcus
inferior lacrimal
canaliculus

superior eyelid
conjunctiva with
Meibomian glands
lacrimal papilla
pupil
iris
nasolacrimal
duct
frontal
maxillary
apophysis
conjunctiva
of sclera
inferior eyelid
conjuctiva with
Meibomian
glands
lacrimal puncta
caruncula lacrimalis

lacrimal sac
medial
palpebral
ligament

superior
lacrimal
canaliculus
eyebrow

lacrimal papilla
superior eyelid
inferior eyelid
lacrimal caruncle
caruncula
lacrimalis
orbicularis
oculi
muscle
inferior
oblique
muscle
maxillary
sinus

4. Sight

DIAGRAM OF VISUAL PATHWAY

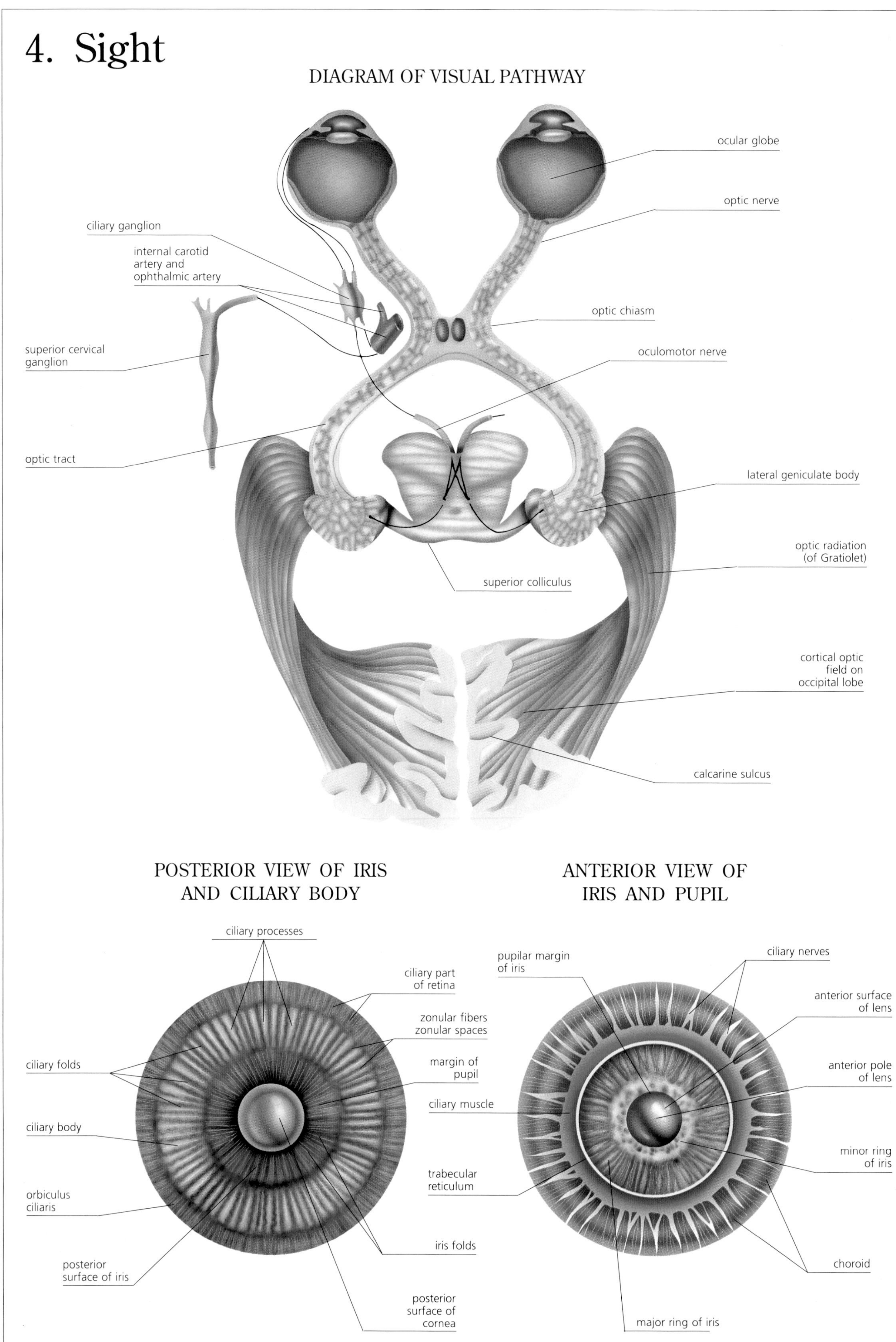

ciliary ganglion

internal carotid
artery and
ophthalmic artery

superior cervical
ganglion

optic tract

ocular globe

optic nerve

optic chiasm

oculomotor nerve

superior colliculus

lateral geniculate body

optic radiation
(of Gratiolet)

cortical optic
field on
occipital lobe

calcarine sulcus

POSTERIOR VIEW OF IRIS
AND CILIARY BODY

ciliary processes

ciliary part
of retina

zonular fibers
zonular spaces

margin of
pupil

ciliary muscle

trabecular
reticulum

iris folds

ciliary folds

ciliary body

orbiculus
ciliaris

posterior
surface of iris

posterior
surface of
cornea

ANTERIOR VIEW OF
IRIS AND PUPIL

pupilar margin
of iris

ciliary nerves

anterior surface
of lens

anterior pole
of lens

minor ring
of iris

choroid

major ring of iris

5. Sight

INFERIOR VIEW OF BRAIN SHOWING THE VISUAL PATHWAYS

olfactory bulb

cornea

ocular nerve

orbital gyri

optic nerve

olfactory tract

optic chiasm

olfactory trigone

optic tract

tuber cinereum

median root

mamillary body

lateral root

midbrain of cerebral peduncle

medial geniculate body

cerebral aqueduct

lateral geniculate body

midbrain top

optic rays

splenium of corpus callosum

brachium of superior colliculus

calcarine sulcus

choroid plexus of lateral ventricle

striated area

MAIN DEFECTS OF SIGHT AND THEIR CORRECTION

PRESBYOPIA

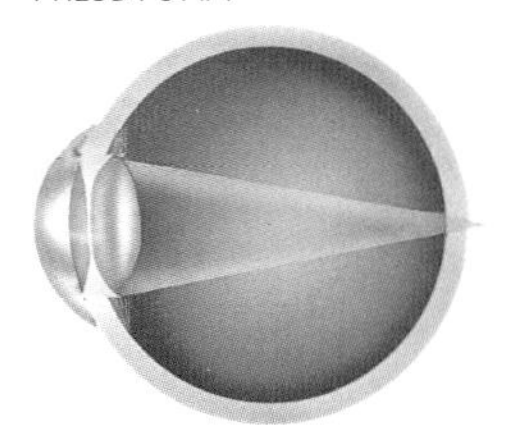

The lens loses elasticity, and does not curve enough. A clear image of close objects forms behind the retina.

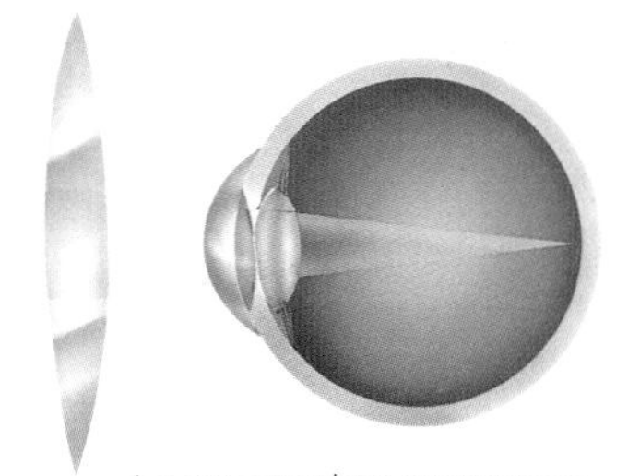

A convergent lens compensates for the lack of angle of the crystalline lens.

MYOPIA (NEAR SIGHTEDNESS)

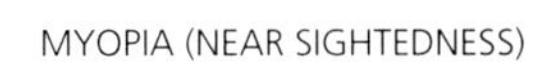

The lens functions correctly, but the ocular globe is too long. A clear image of close objects forms in front of the retina.

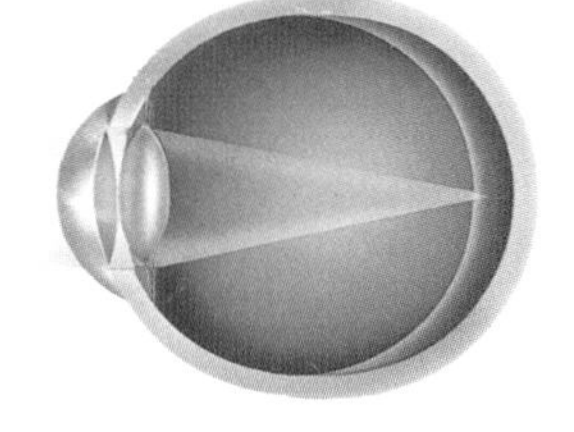

A divergent lens places the clear image on the retina.

HYPERMETROPIA (FAR SIGHTEDNESS)

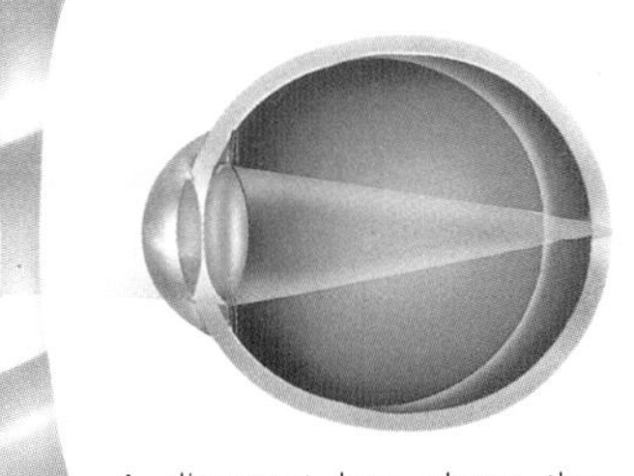

The lens functions correctly, but the ocular globe is too short. A clear image of close objects forms behind the retina.

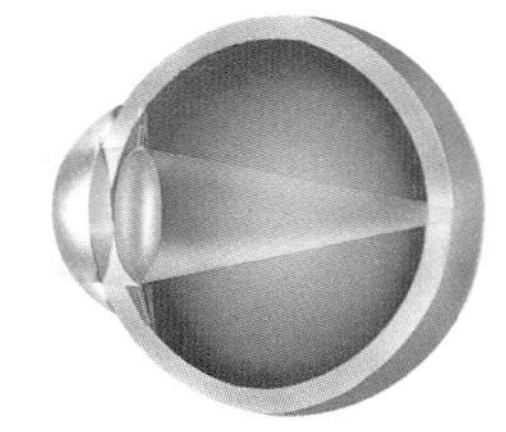

A convergent lens places the clear image on the retina.

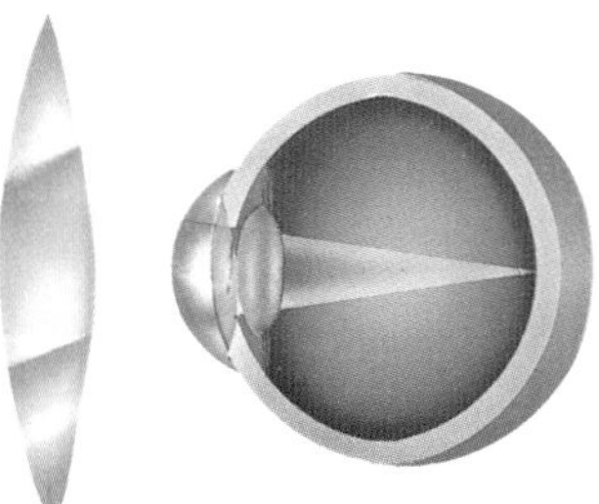

1. Ear

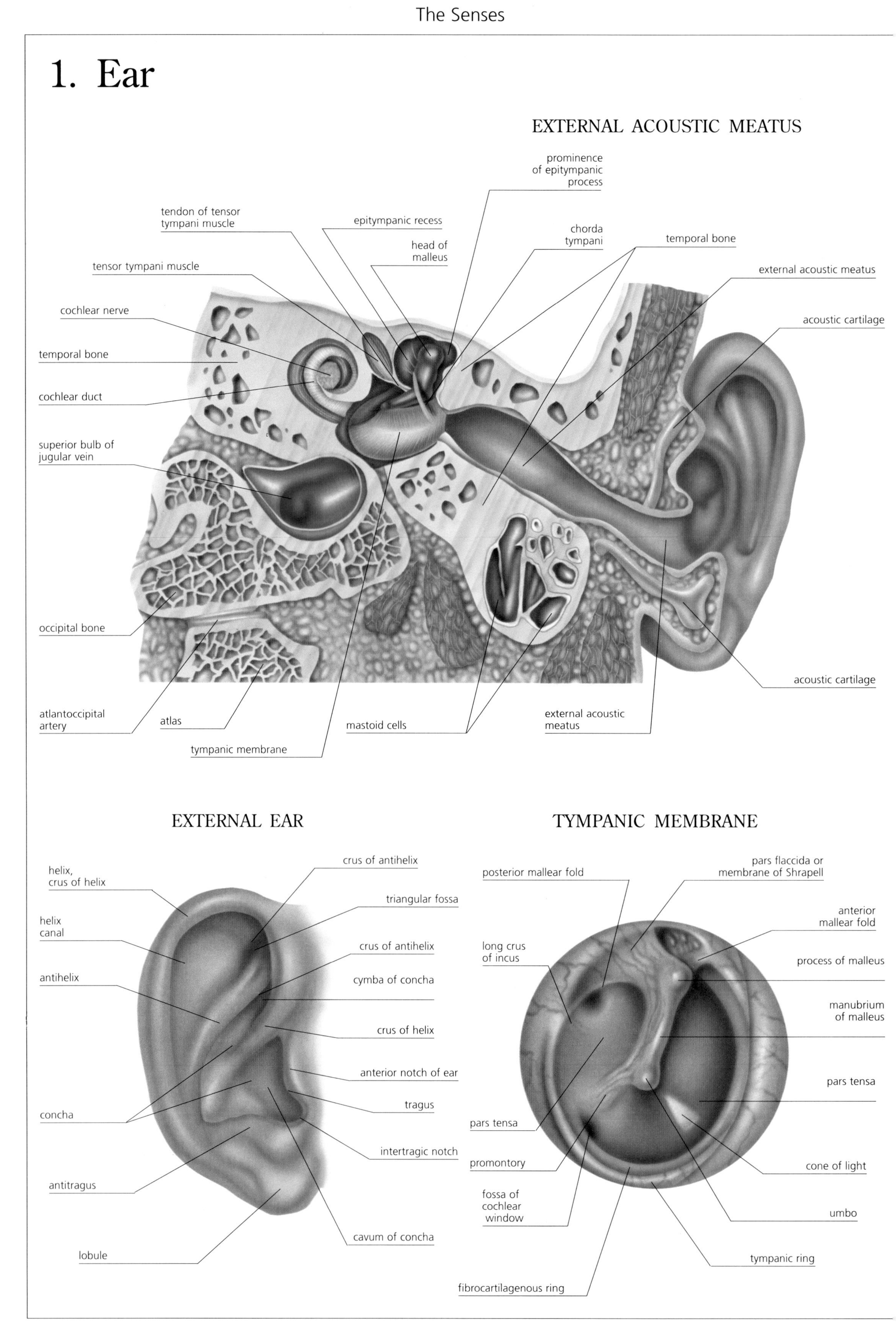

EXTERNAL ACOUSTIC MEATUS

tendon of tensor tympani muscle

prominence of epitympanic process

epitympanic recess

head of malleus

chorda tympani

temporal bone

tensor tympani muscle

external acoustic meatus

cochlear nerve

acoustic cartilage

temporal bone

cochlear duct

superior bulb of jugular vein

occipital bone

acoustic cartilage

atlantoccipital artery

atlas

mastoid cells

external acoustic meatus

tympanic membrane

EXTERNAL EAR

TYMPANIC MEMBRANE

helix, crus of helix

crus of antihelix

posterior mallear fold

pars flaccida or membrane of Shrapell

triangular fossa

anterior mallear fold

helix canal

crus of antihelix

long crus of incus

process of malleus

antihelix

cymba of concha

manubrium of malleus

crus of helix

anterior notch of ear

concha

tragus

pars tensa

pars tensa

antitragus

intertragic notch

promontory

cone of light

cavum of concha

fossa of cochlear window

umbo

lobule

tympanic ring

fibrocartilagenous ring

2. Ear

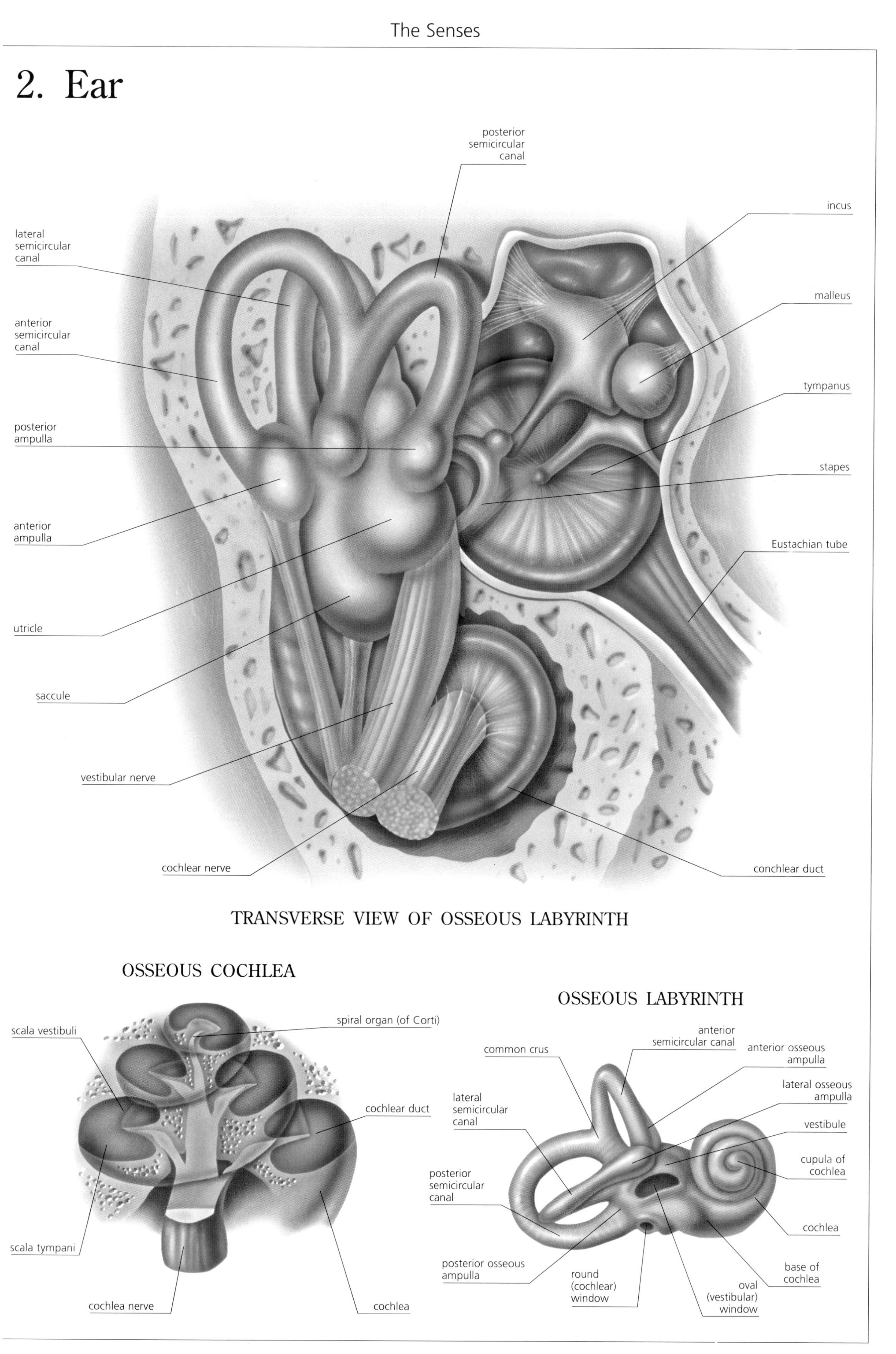

posterior
semicircular
canal

incus

lateral
semicircular
canal

malleus

anterior
semicircular
canal

tympanus

posterior
ampulla

stapes

anterior
ampulla

Eustachian tube

utricle

saccule

vestibular nerve

cochlear nerve

conchlear duct

TRANSVERSE VIEW OF OSSEOUS LABYRINTH

OSSEOUS COCHLEA

spiral organ (of Corti)

scala vestibuli

cochlear duct

scala tympani

cochlea nerve

cochlea

OSSEOUS LABYRINTH

common crus

anterior
semicircular canal

anterior osseous
ampulla

lateral osseous
ampulla

lateral
semicircular
canal

vestibule

cupula of
cochlea

posterior
semicircular
canal

cochlea

posterior osseous
ampulla

round
(cochlear)
window

oval
(vestibular)
window

base of
cochlea

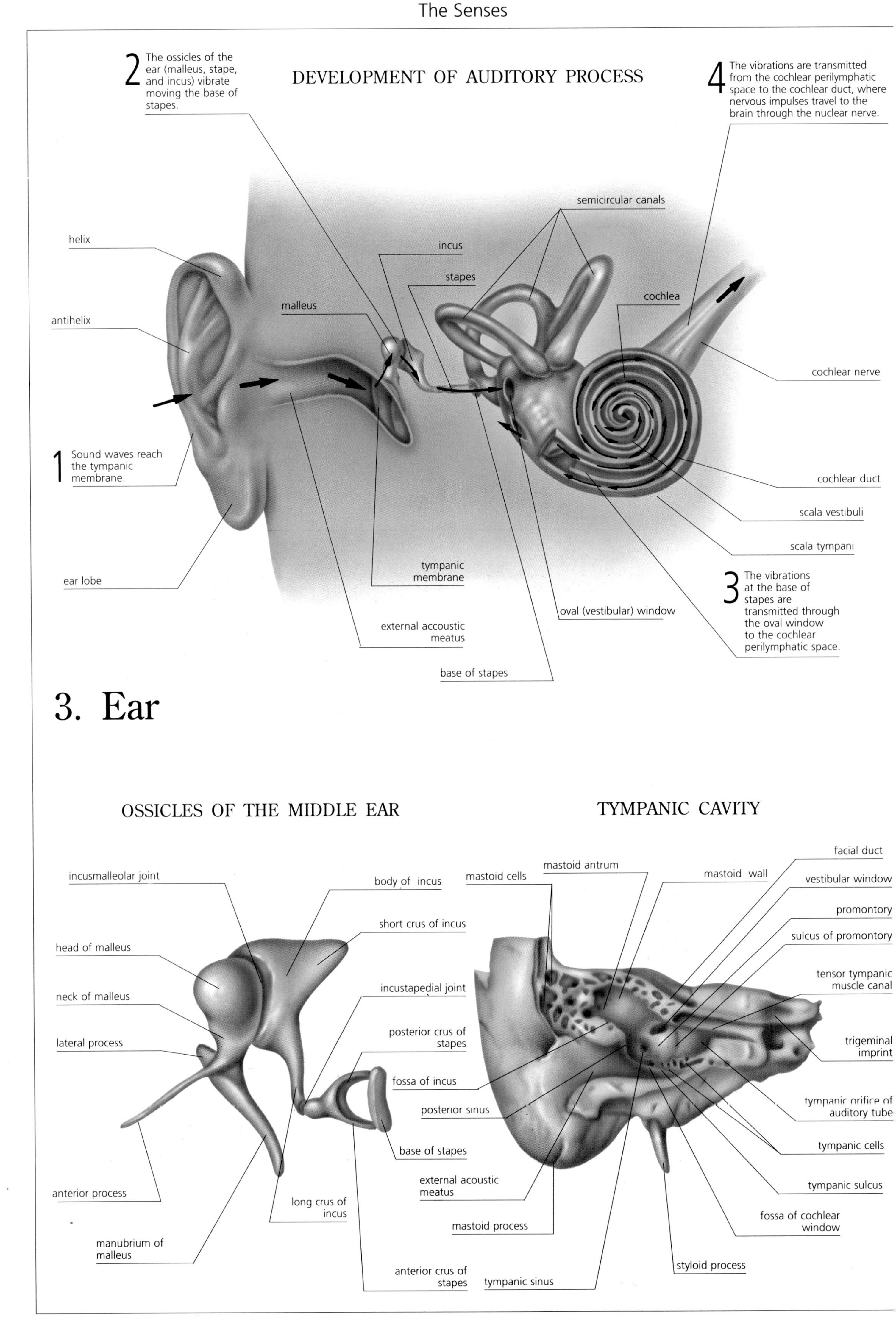

DEVELOPMENT OF AUDITORY PROCESS

2 The ossicles of the ear (malleus, stape, and incus) vibrate moving the base of stapes.

4 The vibrations are transmitted from the cochlear perilymphatic space to the cochlear duct, where nervous impulses travel to the brain through the nuclear nerve.

helix

antihelix

malleus

incus

stapes

semicircular canals

cochlea

cochlear nerve

1 Sound waves reach the tympanic membrane.

ear lobe

tympanic membrane

external accoustic meatus

base of stapes

oval (vestibular) window

cochlear duct

scala vestibuli

scala tympani

3 The vibrations at the base of stapes are transmitted through the oval window to the cochlear perilymphatic space.

3. Ear

OSSICLES OF THE MIDDLE EAR

TYMPANIC CAVITY

incusmalleolar joint

head of malleus

neck of malleus

lateral process

anterior process

manubrium of malleus

body of incus

short crus of incus

incustapedial joint

posterior crus of stapes

fossa of incus

posterior sinus

base of stapes

external acoustic meatus

mastoid process

long crus of incus

anterior crus of stapes

tympanic sinus

mastoid cells

mastoid antrum

mastoid wall

facial duct

vestibular window

promontory

sulcus of promontory

tensor tympanic muscle canal

trigeminal imprint

tympanic orifice of auditory tube

tympanic cells

tympanic sulcus

fossa of cochlear window

styloid process

1. Smell

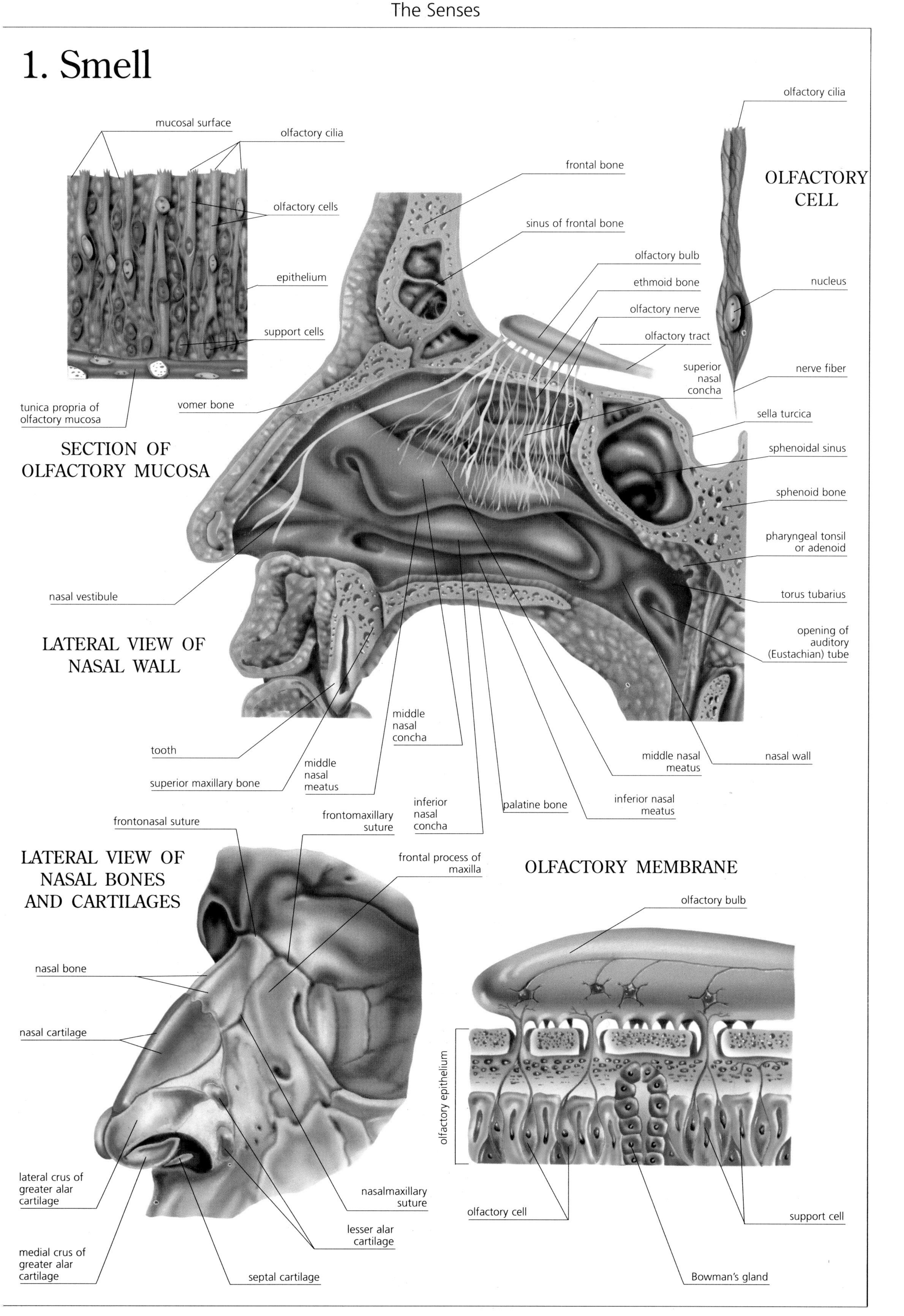

mucosal surface

olfactory cilia

olfactory cells

epithelium

support cells

tunica propria of olfactory mucosa

SECTION OF OLFACTORY MUCOSA

vomer bone

frontal bone

sinus of frontal bone

olfactory bulb

ethmoid bone

olfactory nerve

olfactory tract

superior nasal concha

sella turcica

sphenoidal sinus

sphenoid bone

pharyngeal tonsil or adenoid

torus tubarius

opening of auditory (Eustachian) tube

olfactory cilia

OLFACTORY CELL

nucleus

nerve fiber

nasal vestibule

LATERAL VIEW OF NASAL WALL

tooth

superior maxillary bone

middle nasal meatus

middle nasal concha

inferior nasal concha

palatine bone

middle nasal meatus

nasal wall

inferior nasal meatus

frontonasal suture

frontomaxillary suture

frontal process of maxilla

LATERAL VIEW OF NASAL BONES AND CARTILAGES

OLFACTORY MEMBRANE

olfactory bulb

nasal bone

nasal cartilage

olfactory epithelium

lateral crus of greater alar cartilage

nasalmaxillary suture

medial crus of greater alar cartilage

lesser alar cartilage

septal cartilage

olfactory cell

support cell

Bowman's gland

1. Taste

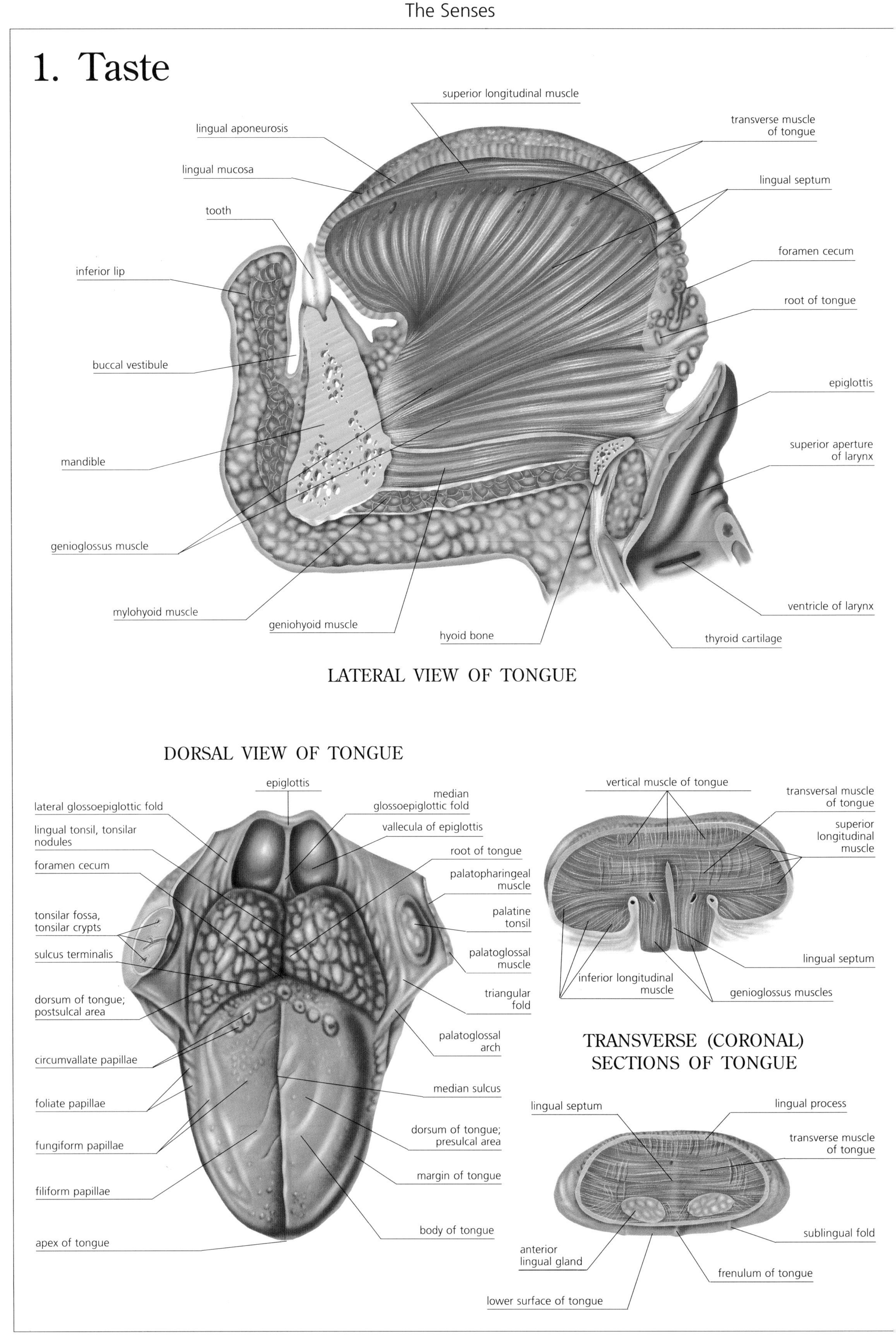

superior longitudinal muscle

lingual aponeurosis

lingual mucosa

tooth

inferior lip

buccal vestibule

mandible

genioglossus muscle

mylohyoid muscle

geniohyoid muscle

hyoid bone

transverse muscle of tongue

lingual septum

foramen cecum

root of tongue

epiglottis

superior aperture of larynx

ventricle of larynx

thyroid cartilage

LATERAL VIEW OF TONGUE

DORSAL VIEW OF TONGUE

lateral glossoepiglottic fold

lingual tonsil, tonsilar nodules

foramen cecum

tonsilar fossa, tonsilar crypts

sulcus terminalis

dorsum of tongue; postsulcal area

circumvallate papillae

foliate papillae

fungiform papillae

filiform papillae

apex of tongue

epiglottis

median glossoepiglottic fold

vallecula of epiglottis

root of tongue

palatopharingeal muscle

palatine tonsil

palatoglossal muscle

triangular fold

palatoglossal arch

median sulcus

dorsum of tongue; presulcal area

margin of tongue

body of tongue

vertical muscle of tongue

transversal muscle of tongue

superior longitudinal muscle

inferior longitudinal muscle

lingual septum

genioglossus muscles

TRANSVERSE (CORONAL)
SECTIONS OF TONGUE

lingual septum

lingual process

transverse muscle of tongue

sublingual fold

anterior lingual gland

frenulum of tongue

lower surface of tongue

2. Taste

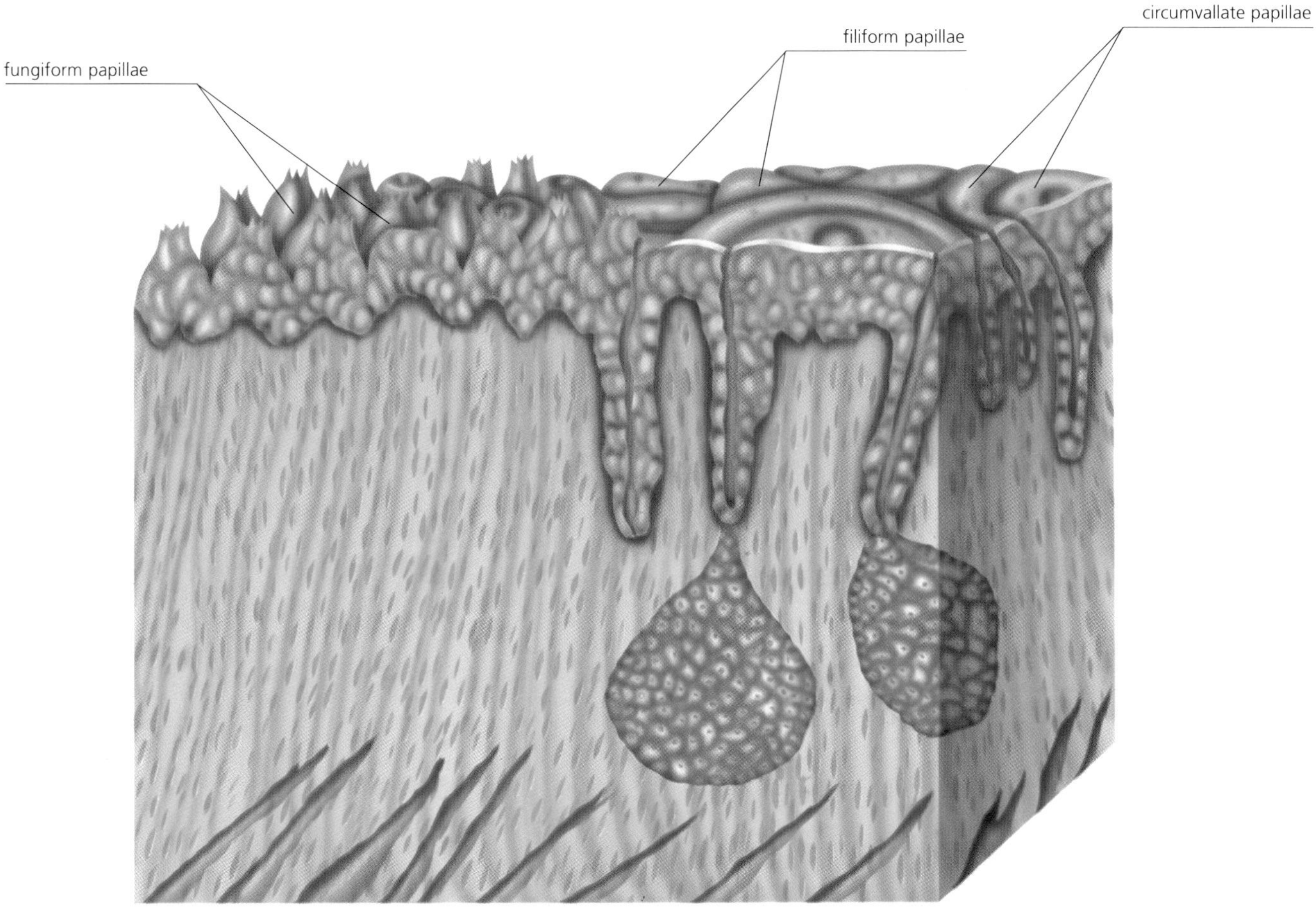

fungiform papillae

filiform papillae

circumvallate papillae

SCHEMATIC SECTION OF PARTS OF THE TONGUE

SUPERIOR VIEW OF TONGUE
SHOWING TASTE AREAS

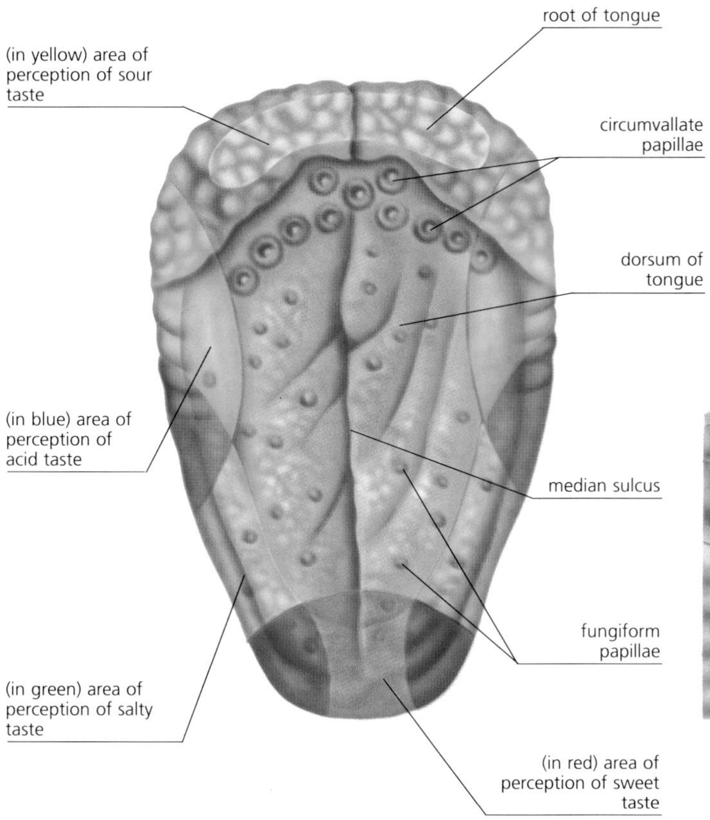

(in yellow) area of
perception of sour
taste

root of tongue

circumvallate
papillae

dorsum of
tongue

(in blue) area of
perception of
acid taste

median sulcus

fungiform
papillae

(in green) area of
perception of salty
taste

(in red) area of
perception of sweet
taste

taste villi

taste cells

SCHEMATIC SECTION
OF PART OF A
CIRCUMVALLATE
PAPILLA

nerves

taste bud

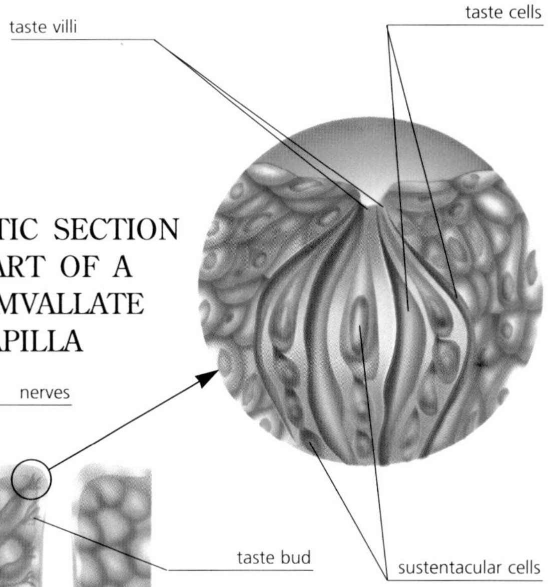

sustentacular cells

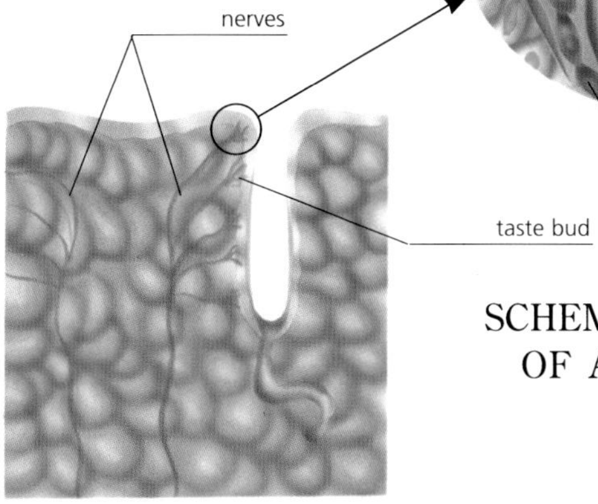

SCHEMATIC SECTION
OF A TASTE BUD

3. Taste

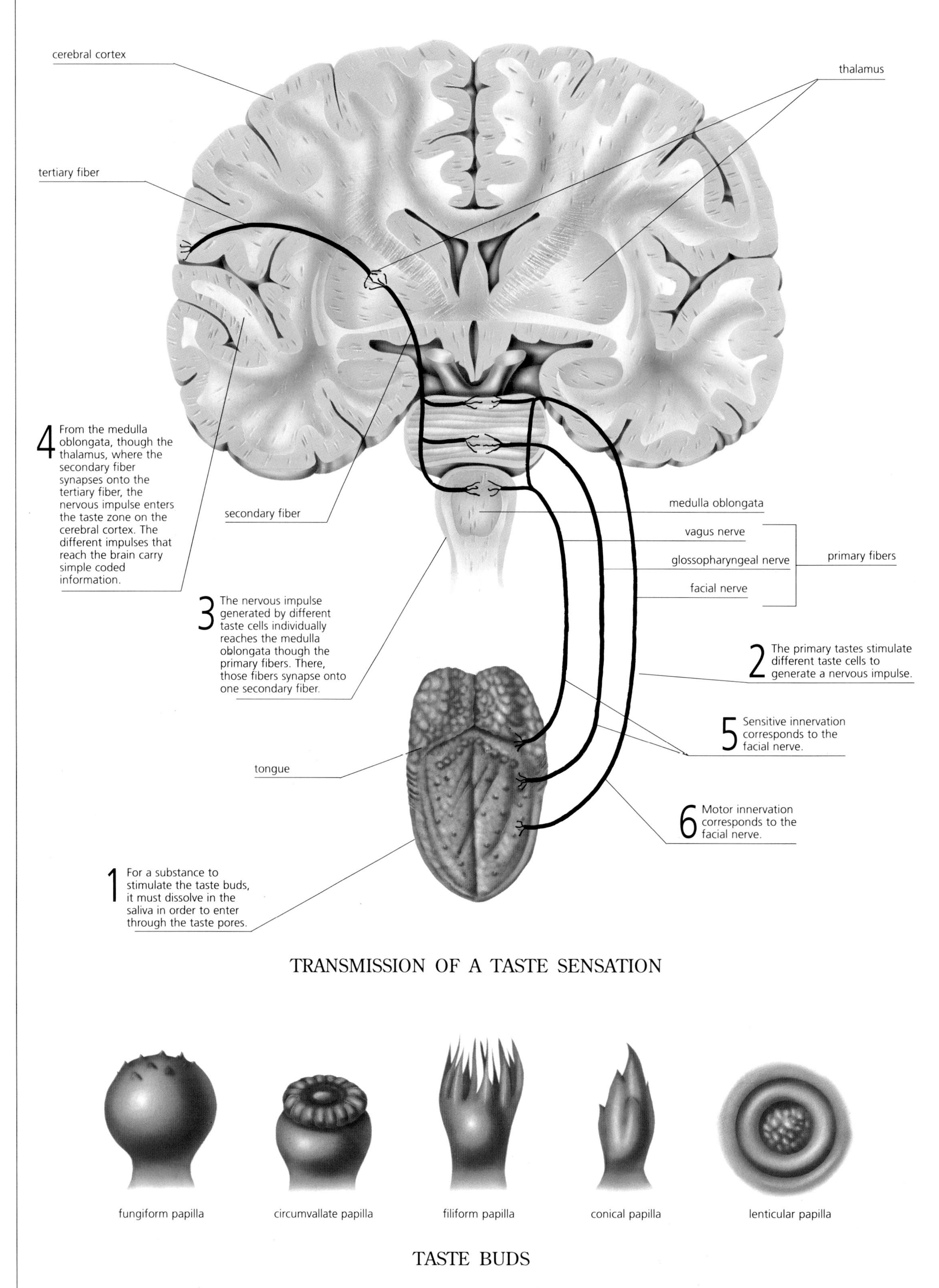

cerebral cortex

thalamus

tertiary fiber

4 From the medulla oblongata, though the thalamus, where the secondary fiber synapses onto the tertiary fiber, the nervous impulse enters the taste zone on the cerebral cortex. The different impulses that reach the brain carry simple coded information.

secondary fiber

3 The nervous impulse generated by different taste cells individually reaches the medulla oblongata though the primary fibers. There, those fibers synapse onto one secondary fiber.

medulla oblongata

vagus nerve

glossopharyngeal nerve

facial nerve

primary fibers

2 The primary tastes stimulate different taste cells to generate a nervous impulse.

5 Sensitive innervation corresponds to the facial nerve.

tongue

6 Motor innervation corresponds to the facial nerve.

1 For a substance to stimulate the taste buds, it must dissolve in the saliva in order to enter through the taste pores.

TRANSMISSION OF A TASTE SENSATION

fungiform papilla

circumvallate papilla

filiform papilla

conical papilla

lenticular papilla

TASTE BUDS

1. Touch

SCHEMATIC SECTION
OF THE SKIN

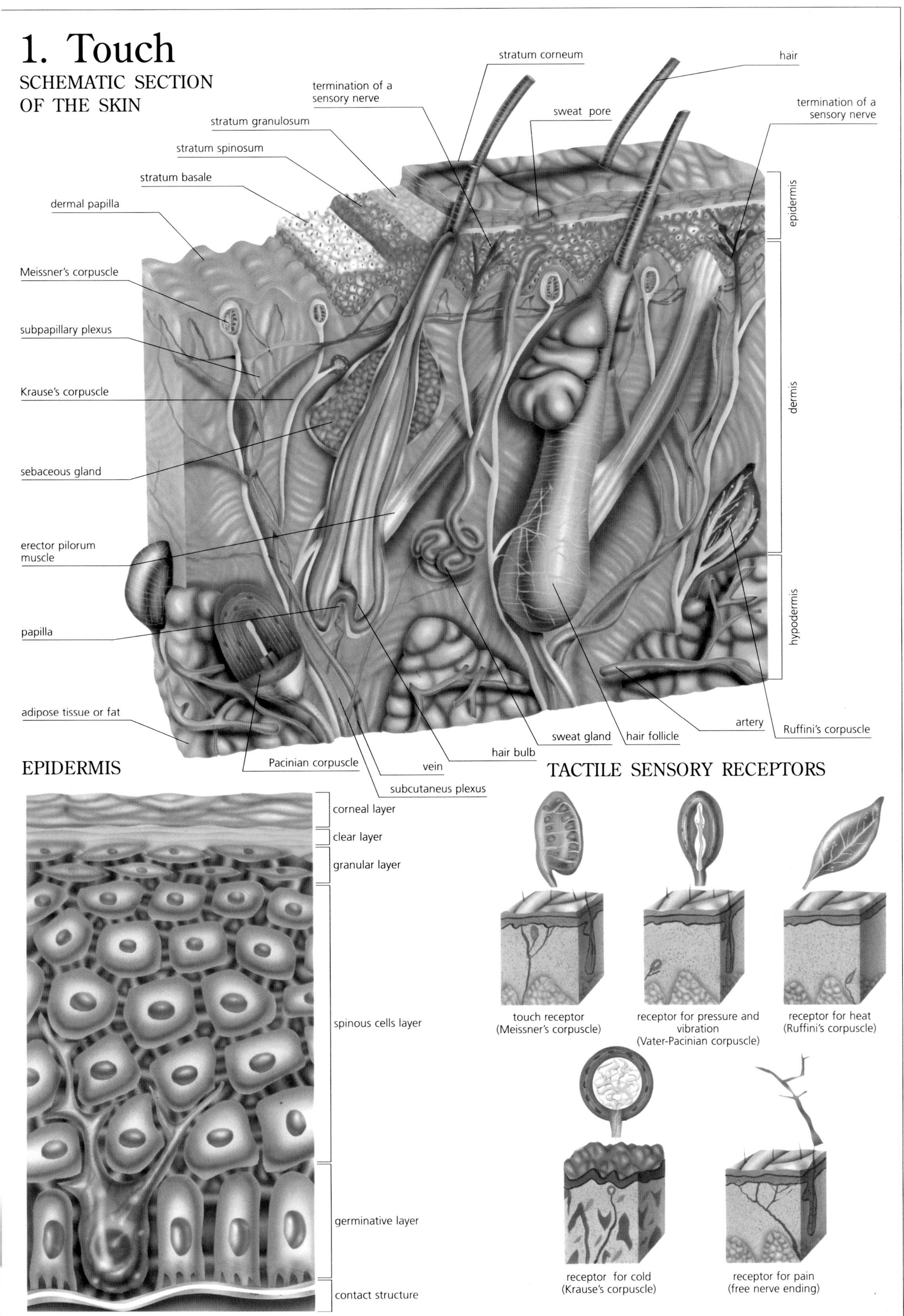

stratum corneum

hair

termination of a
sensory nerve

sweat pore

termination of a
sensory nerve

stratum granulosum

stratum spinosum

stratum basale

dermal papilla

epidermis

Meissner's corpuscle

subpapillary plexus

Krause's corpuscle

dermis

sebaceous gland

erector pilorum
muscle

papilla

hypodermis

adipose tissue or fat

artery

Ruffini's corpuscle

sweat gland

hair follicle

Pacinian corpuscle

vein

hair bulb

subcutaneus plexus

EPIDERMIS

corneal layer

clear layer

granular layer

spinous cells layer

germinative layer

contact structure

TACTILE SENSORY RECEPTORS

touch receptor
(Meissner's corpuscle)

receptor for pressure and
vibration
(Vater-Pacinian corpuscle)

receptor for heat
(Ruffini's corpuscle)

receptor for cold
(Krause's corpuscle)

receptor for pain
(free nerve ending)

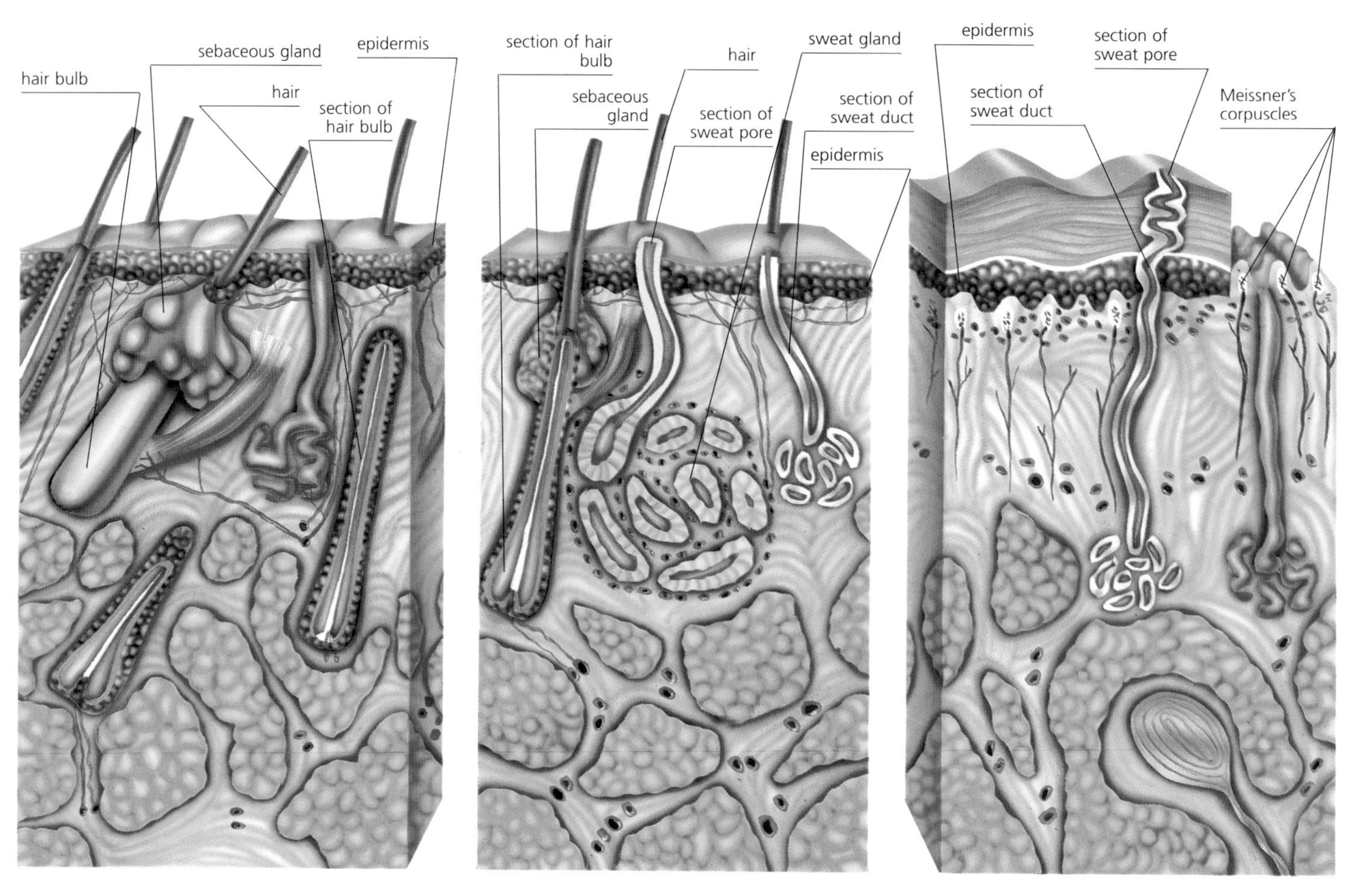

hair bulb

sebaceous gland

hair

epidermis

section of hair bulb

section of hair bulb

sebaceous gland

hair

sweat gland

section of sweat pore

section of sweat duct

epidermis

epidermis

section of sweat duct

section of sweat pore

Meissner's corpuscles

SCALP

AXILLA

SOLE OF FOOT

SCHEMATIC SECTIONS OF DIFFERENT TYPES OF SKIN OF THE HUMAN BODY

SCHEMATIC SECTION OF A HAIR

2. Touch

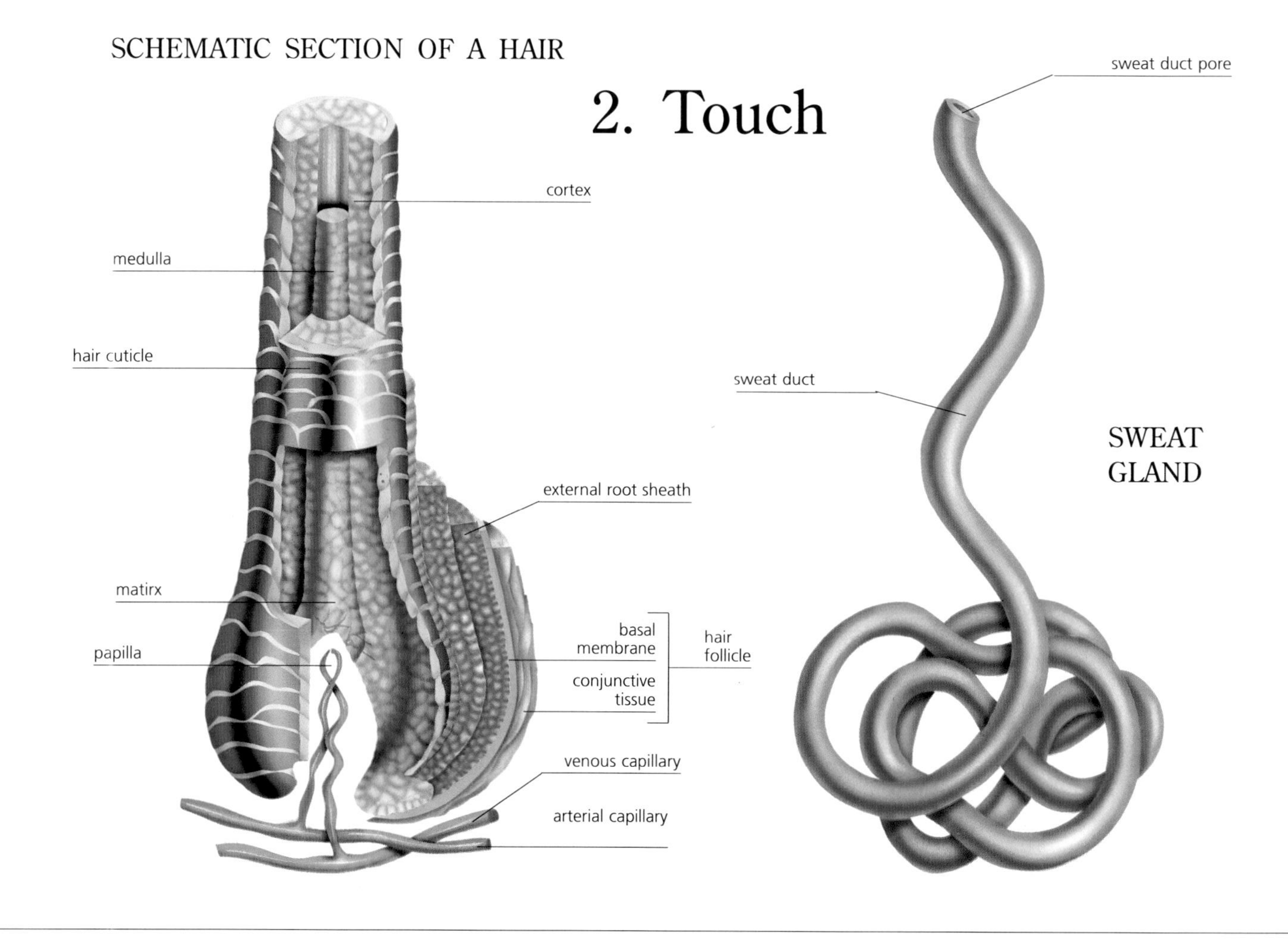

cortex

medulla

hair cuticle

matirx

papilla

external root sheath

basal membrane

hair follicle

conjunctive tissue

venous capillary

arterial capillary

sweat duct pore

sweat duct

SWEAT GLAND

3. Touch

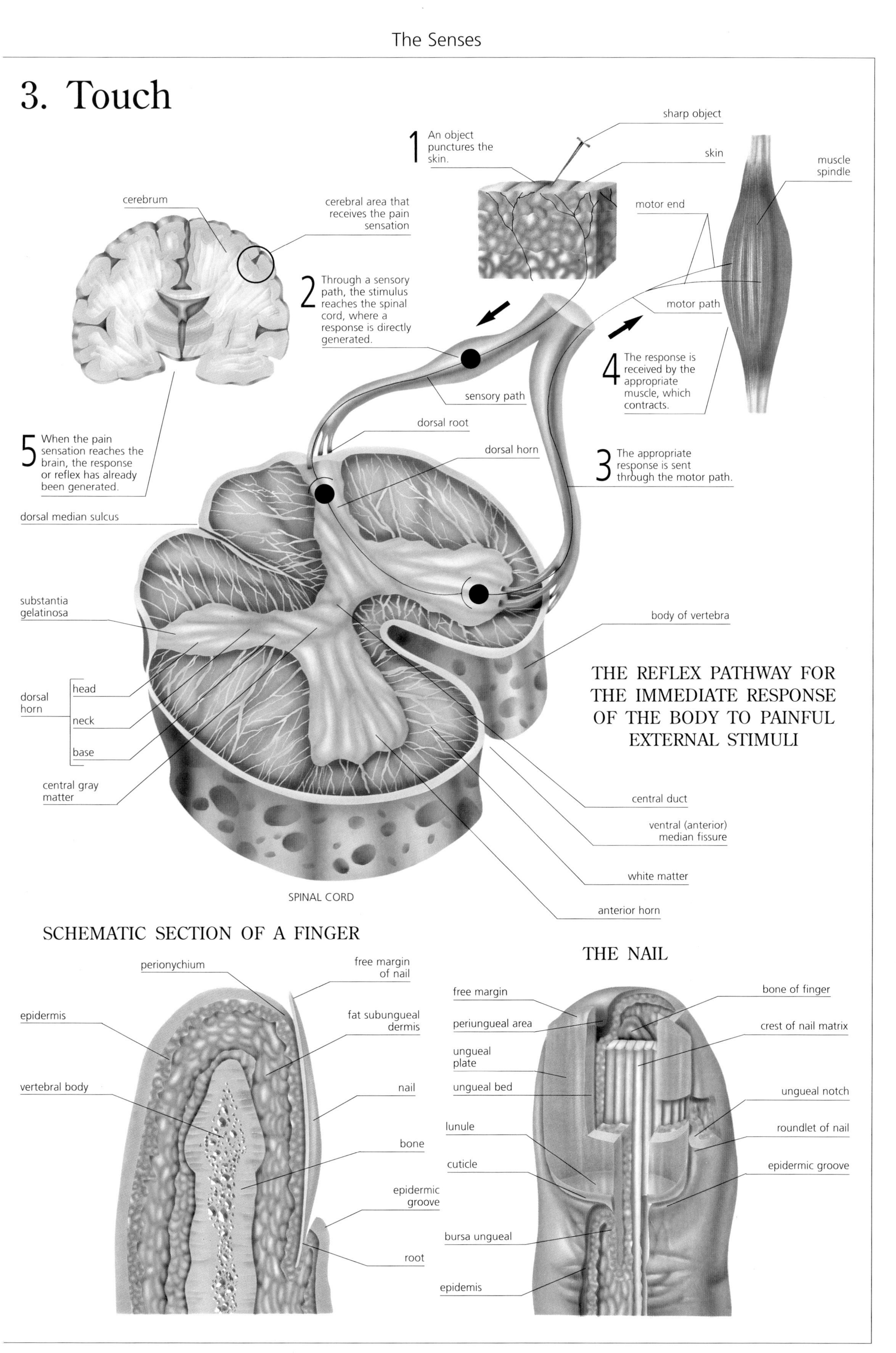

1 An object punctures the skin.

sharp object

skin

muscle spindle

cerebrum

cerebral area that receives the pain sensation

motor end

2 Through a sensory path, the stimulus reaches the spinal cord, where a response is directly generated.

motor path

sensory path

4 The response is received by the appropriate muscle, which contracts.

dorsal root

dorsal horn

5 When the pain sensation reaches the brain, the response or reflex has already been generated.

3 The appropriate response is sent through the motor path.

dorsal median sulcus

substantia gelatinosa

body of vertebra

dorsal horn — head — neck — base

central gray matter

THE REFLEX PATHWAY FOR THE IMMEDIATE RESPONSE OF THE BODY TO PAINFUL EXTERNAL STIMULI

central duct

ventral (anterior) median fissure

white matter

SPINAL CORD

anterior horn

SCHEMATIC SECTION OF A FINGER

perionychium

free margin of nail

epidermis

fat subungueal dermis

vertebral body

nail

bone

epidermic groove

root

THE NAIL

free margin

bone of finger

periungueal area

crest of nail matrix

ungueal plate

ungueal bed

ungueal notch

lunule

roundlet of nail

cuticle

epidermic groove

bursa ungueal

epidemis

The Urinary System
1. Male Urinary System.
Female Urinary System

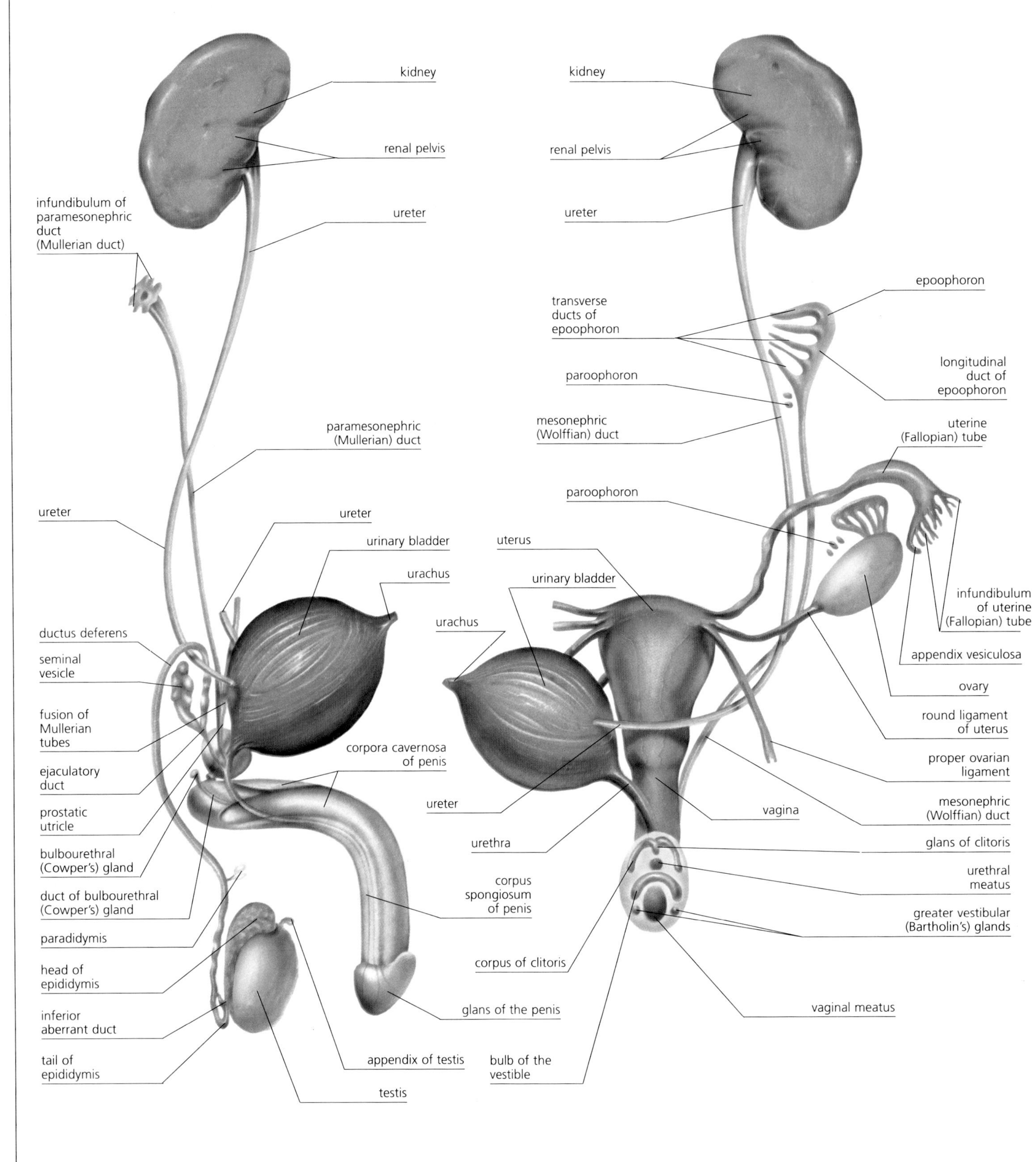

kidney

renal pelvis

ureter

infundibulum of
paramesonephric
duct
(Mullerian duct)

paramesonephric
(Mullerian) duct

ureter

ureter

urinary bladder

urachus

ductus deferens

seminal
vesicle

fusion of
Mullerian
tubes

ejaculatory
duct

corpora cavernosa
of penis

prostatic
utricle

bulbourethral
(Cowper's) gland

duct of bulbourethral
(Cowper's) gland

paradidymis

head of
epididymis

inferior
aberrant duct

tail of
epididymis

appendix of testis

testis

corpus
spongiosum
of penis

corpus of clitoris

glans of the penis

bulb of the
vestible

kidney

renal pelvis

ureter

transverse
ducts of
epoophoron

paroophoron

mesonephric
(Wolffian) duct

paroophoron

uterus

urinary bladder

urachus

ureter

urethra

epoophoron

longitudinal
duct of
epoophoron

uterine
(Fallopian) tube

infundibulum
of uterine
(Fallopian) tube

appendix vesiculosa

ovary

round ligament
of uterus

proper ovarian
ligament

mesonephric
(Wolffian) duct

glans of clitoris

urethral
meatus

greater vestibular
(Bartholin's) glands

vagina

vaginal meatus

LATERAL VIEW OF MALE
GENITOURINARY SYSTEM

LATERAL VIEW OF FEMALE
GENITOURINARY SYSTEM

2. Kidney

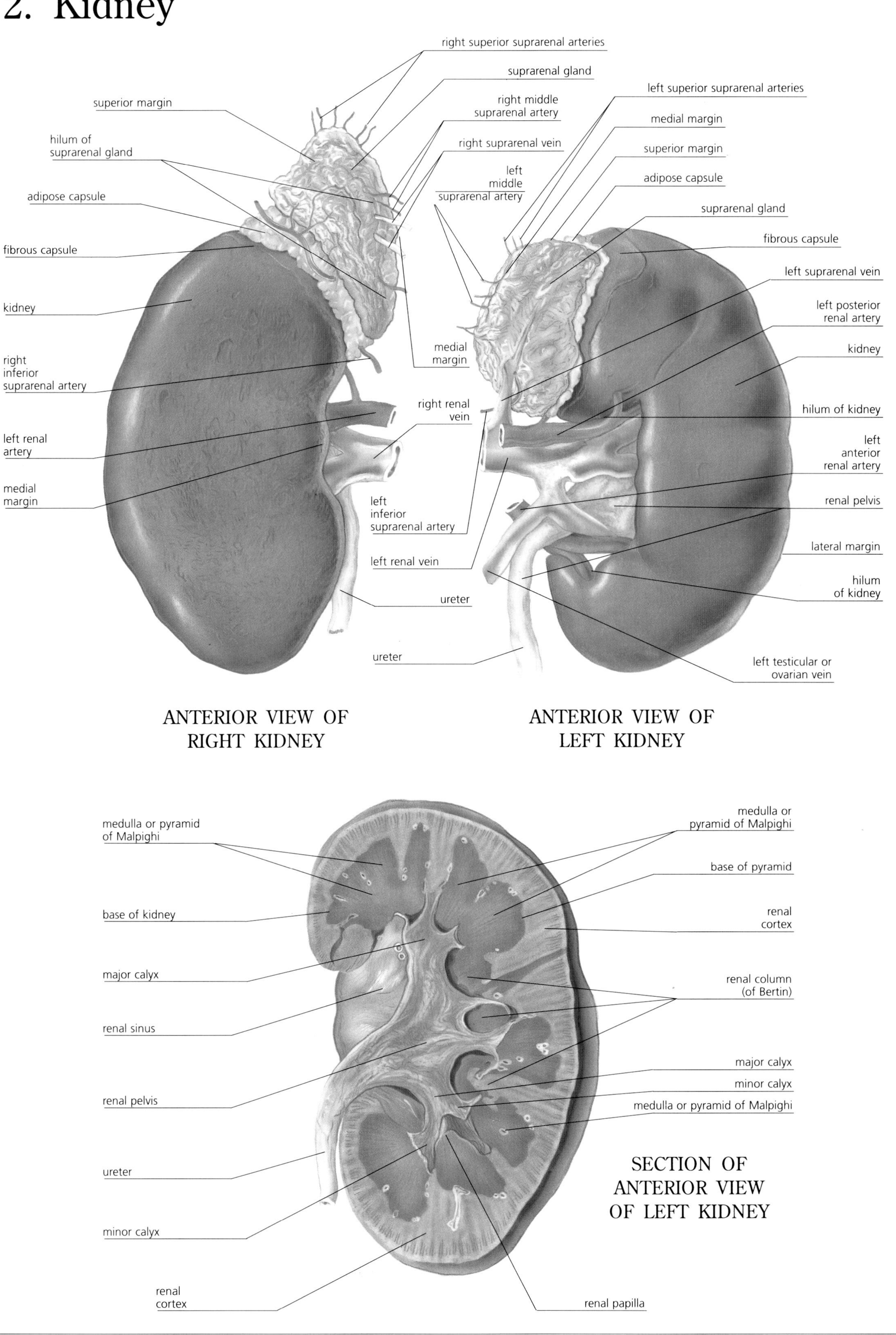

right superior suprarenal arteries

suprarenal gland

right middle
suprarenal artery

left superior suprarenal arteries

medial margin

superior margin

right suprarenal vein

adipose capsule

left
middle
suprarenal artery

suprarenal gland

fibrous capsule

superior margin

hilum of
suprarenal gland

left suprarenal vein

adipose capsule

left posterior
renal artery

fibrous capsule

kidney

kidney

hilum of kidney

right
inferior
suprarenal artery

medial
margin

right renal
vein

left
anterior
renal artery

left renal
artery

renal pelvis

medial
margin

left
inferior
suprarenal artery

lateral margin

hilum
of kidney

left renal vein

ureter

left testicular or
ovarian vein

ureter

ANTERIOR VIEW OF
RIGHT KIDNEY

ANTERIOR VIEW OF
LEFT KIDNEY

medulla or pyramid
of Malpighi

medulla or
pyramid of Malpighi

base of pyramid

base of kidney

renal
cortex

major calyx

renal column
(of Bertin)

renal sinus

major calyx

renal pelvis

minor calyx

medulla or pyramid of Malpighi

ureter

SECTION OF
ANTERIOR VIEW
OF LEFT KIDNEY

minor calyx

renal
cortex

renal papilla

75

3. Nephron and Glomerulus

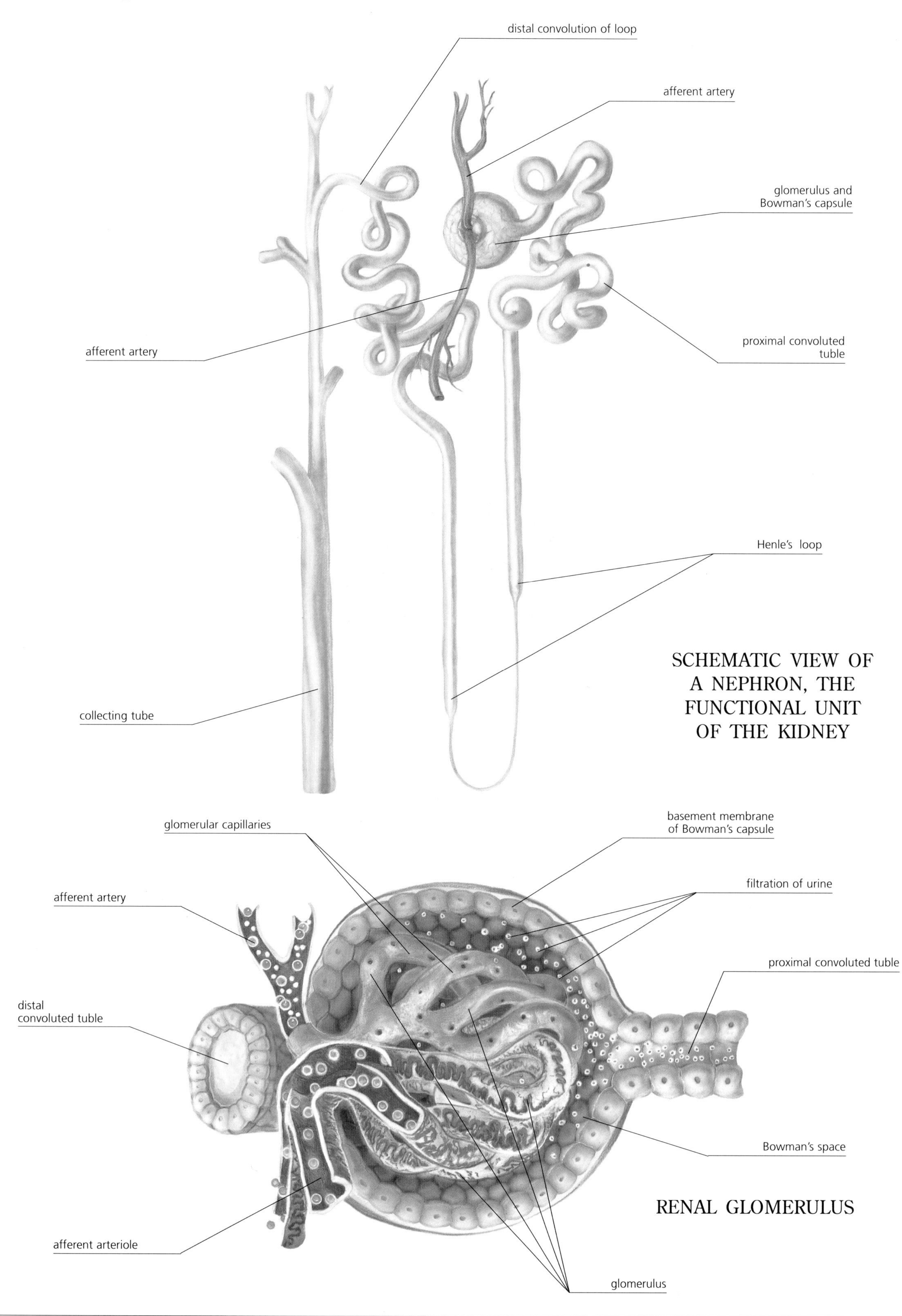

distal convolution of loop

afferent artery

glomerulus and
Bowman's capsule

proximal convoluted
tuble

afferent artery

Henle's loop

collecting tube

SCHEMATIC VIEW OF
A NEPHRON, THE
FUNCTIONAL UNIT
OF THE KIDNEY

glomerular capillaries

basement membrane
of Bowman's capsule

filtration of urine

afferent artery

proximal convoluted tubule

distal
convoluted tuble

Bowman's space

afferent arteriole

RENAL GLOMERULUS

glomerulus

4. Urinary Bladder and Urethra

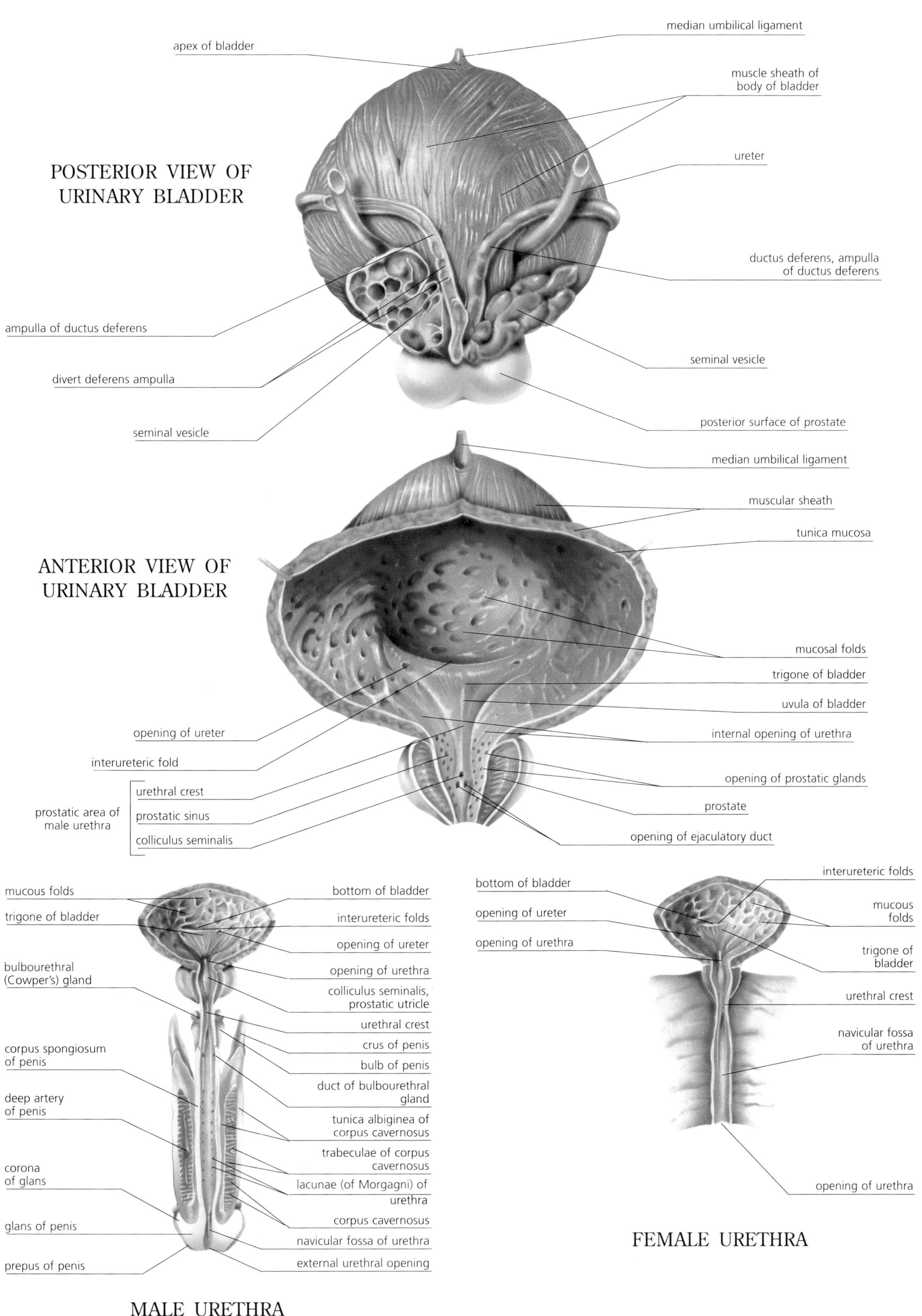

POSTERIOR VIEW OF
URINARY BLADDER

apex of bladder

median umbilical ligament

muscle sheath of
body of bladder

ureter

ductus deferens, ampulla
of ductus deferens

ampulla of ductus deferens

divert deferens ampulla

seminal vesicle

seminal vesicle

posterior surface of prostate

median umbilical ligament

muscular sheath

tunica mucosa

ANTERIOR VIEW OF
URINARY BLADDER

mucosal folds

trigone of bladder

uvula of bladder

opening of ureter

internal opening of urethra

interureteric fold

opening of prostatic glands

prostatic area of
male urethra

urethral crest

prostatic sinus

prostate

colliculus seminalis

opening of ejaculatory duct

mucous folds

trigone of bladder

bottom of bladder

interureteric folds

opening of ureter

bulbourethral
(Cowper's) gland

opening of urethra

colliculus seminalis,
prostatic utricle

urethral crest

corpus spongiosum
of penis

crus of penis

bulb of penis

deep artery
of penis

duct of bulbourethral
gland

tunica albiginea of
corpus cavernosus

trabeculae of corpus
cavernosus

corona
of glans

lacunae (of Morgagni) of
urethra

corpus cavernosus

glans of penis

navicular fossa of urethra

prepus of penis

external urethral opening

MALE URETHRA

bottom of bladder

interureteric folds

opening of ureter

mucous
folds

opening of urethra

trigone of
bladder

urethral crest

navicular fossa
of urethra

opening of urethra

FEMALE URETHRA

77

The Reproductive System
1. Male Reproductive Organs

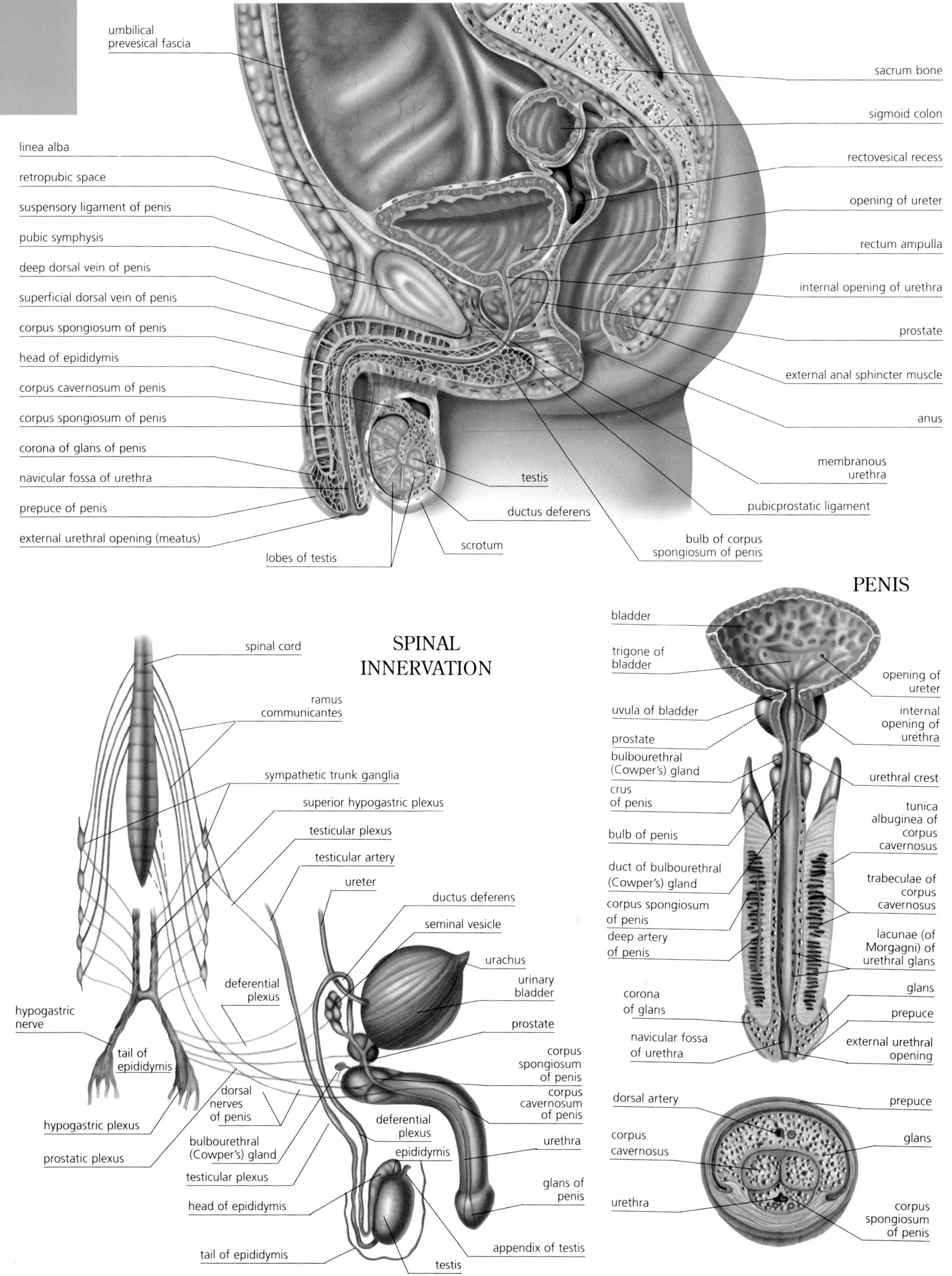

umbilical prevesical fascia

sacrum bone

sigmoid colon

rectovesical recess

opening of ureter

rectum ampulla

internal opening of urethra

prostate

external anal sphincter muscle

anus

membranous urethra

pubicprostatic ligament

linea alba

retropubic space

suspensory ligament of penis

pubic symphysis

deep dorsal vein of penis

superficial dorsal vein of penis

corpus spongiosum of penis

head of epididymis

corpus cavernosum of penis

corpus spongiosum of penis

corona of glans of penis

navicular fossa of urethra

prepuce of penis

external urethral opening (meatus)

testis

ductus deferens

scrotum

bulb of corpus spongiosum of penis

lobes of testis

SPINAL INNERVATION

spinal cord

ramus communicantes

sympathetic trunk ganglia

superior hypogastric plexus

testicular plexus

testicular artery

ureter

ductus deferens

seminal vesicle

urachus

urinary bladder

prostate

corpus spongiosum of penis

corpus cavernosum of penis

urethra

glans of penis

appendix of testis

deferential plexus

hypogastric nerve

tail of epididymis

dorsal nerves of penis

hypogastric plexus

prostatic plexus

bulbourethral (Cowper's) gland

testicular plexus

head of epididymis

tail of epididymis

deferential plexus

epididymis

testis

PENIS

bladder

trigone of bladder

uvula of bladder

prostate

bulbourethral (Cowper's) gland

crus of penis

bulb of penis

duct of bulbourethral (Cowper's) gland

corpus spongiosum of penis

deep artery of penis

corona of glans

navicular fossa of urethra

dorsal artery

corpus cavernosus

urethra

opening of ureter

internal opening of urethra

urethral crest

tunica albuginea of corpus cavernosus

trabeculae of corpus cavernosus

lacunae (of Morgagni) of urethral glans

glans

prepuce

external urethral opening

prepuce

glans

corpus spongiosum of penis

2. Female Reproductive Organs

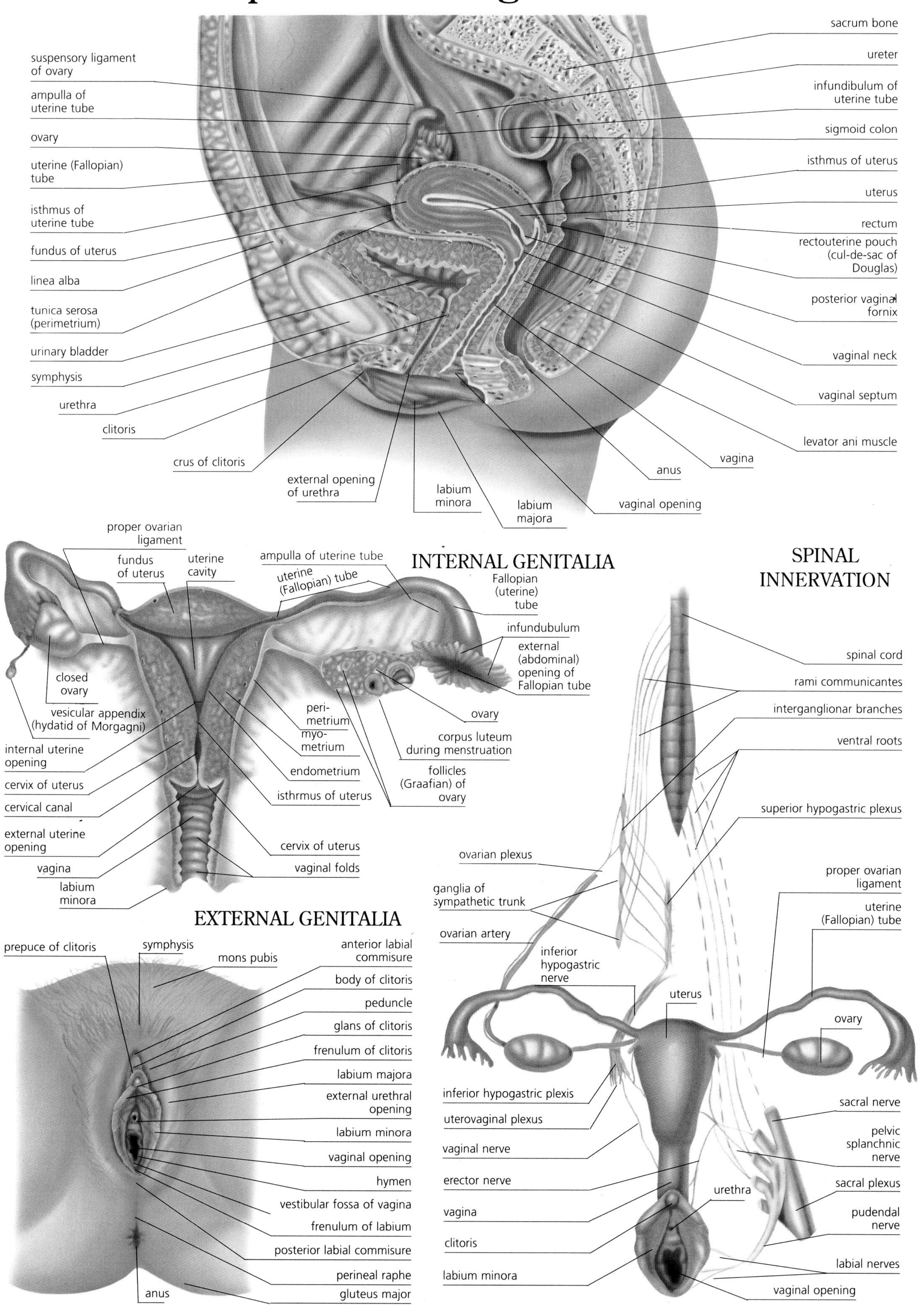

suspensory ligament of ovary

ampulla of uterine tube

ovary

uterine (Fallopian) tube

isthmus of uterine tube

fundus of uterus

linea alba

tunica serosa (perimetrium)

urinary bladder

symphysis

urethra

clitoris

crus of clitoris

external opening of urethra

labium minora

labium majora

vaginal opening

anus

vagina

sacrum bone

ureter

infundibulum of uterine tube

sigmoid colon

isthmus of uterus

uterus

rectum

rectouterine pouch (cul-de-sac of Douglas)

posterior vaginal fornix

vaginal neck

vaginal septum

levator ani muscle

INTERNAL GENITALIA

proper ovarian ligament

fundus of uterus

uterine cavity

ampulla of uterine tube

uterine (Fallopian) tube

closed ovary

vesicular appendix (hydatid of Morgagni)

internal uterine opening

cervix of uterus

cervical canal

external uterine opening

vagina

labium minora

peri-metrium

myo-metrium

endometrium

isthrmus of uterus

cervix of uterus

vaginal folds

Fallopian (uterine) tube

infundubulum

external (abdominal) opening of Fallopian tube

ovary

corpus luteum during menstruation

follicles (Graafian) of ovary

EXTERNAL GENITALIA

prepuce of clitoris

symphysis

mons pubis

anterior labial commisure

body of clitoris

peduncle

glans of clitoris

frenulum of clitoris

labium majora

external urethral opening

labium minora

vaginal opening

hymen

vestibular fossa of vagina

frenulum of labium

posterior labial commisure

perineal raphe

gluteus major

anus

SPINAL INNERVATION

spinal cord

rami communicantes

interganglionar branches

ventral roots

superior hypogastric plexus

proper ovarian ligament

uterine (Fallopian) tube

ovary

ovarian plexus

ganglia of sympathetic trunk

ovarian artery

inferior hypogastric nerve

uterus

inferior hypogastric plexis

uterovaginal plexus

vaginal nerve

erector nerve

vagina

clitoris

labium minora

urethra

sacral nerve

pelvic splanchnic nerve

sacral plexus

pudendal nerve

labial nerves

vaginal opening

3. Male Reproductive Elements

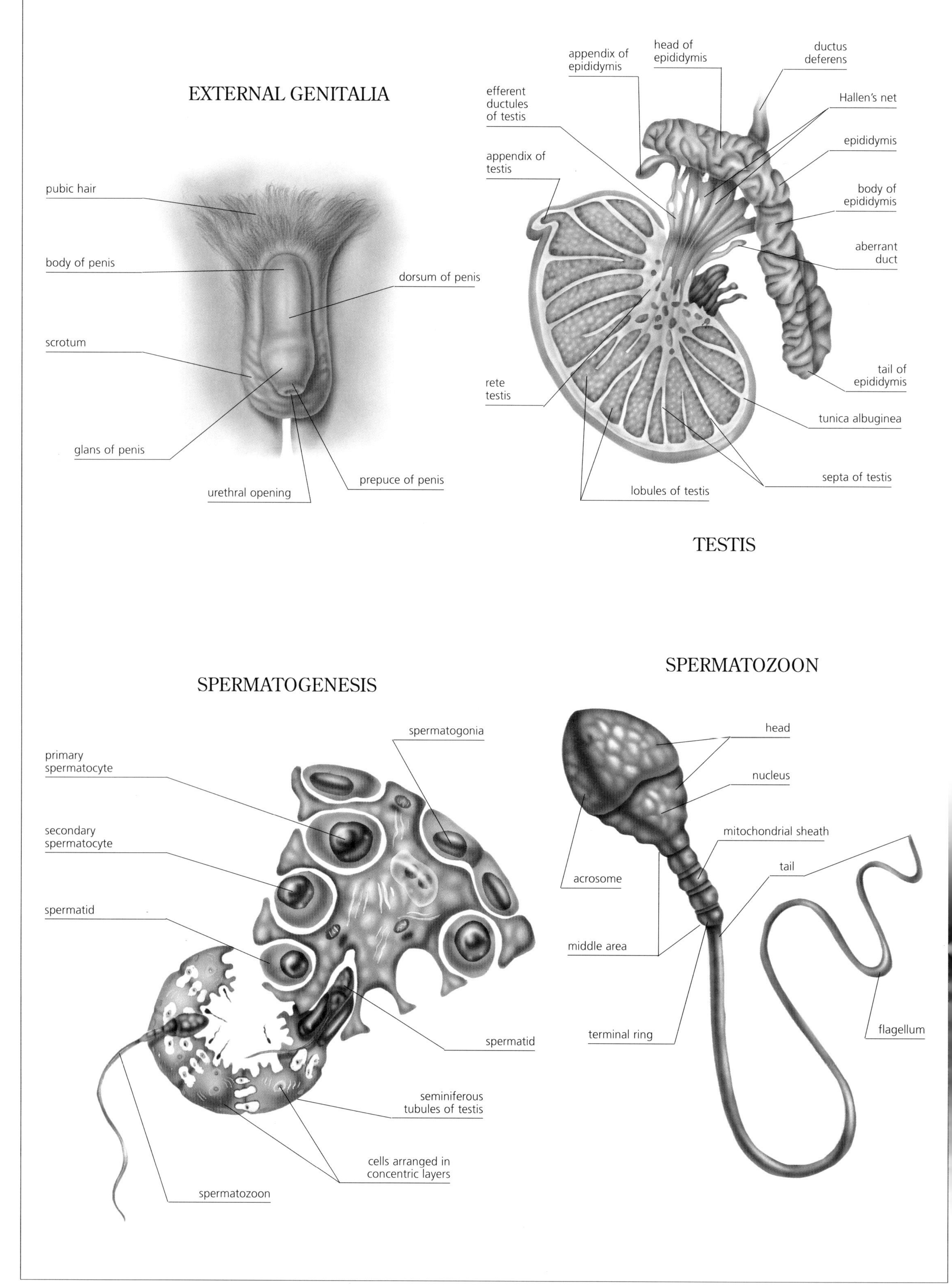

EXTERNAL GENITALIA

pubic hair

body of penis

scrotum

glans of penis

urethral opening

prepuce of penis

dorsum of penis

appendix of epididymis

head of epididymis

ductus deferens

efferent ductules of testis

Hallen's net

epididymis

appendix of testis

body of epididymis

aberrant duct

rete testis

tail of epididymis

tunica albuginea

lobules of testis

septa of testis

TESTIS

SPERMATOGENESIS

SPERMATOZOON

spermatogonia

head

nucleus

primary spermatocyte

secondary spermatocyte

mitochondrial sheath

tail

spermatid

acrosome

spermatid

middle area

seminiferous tubules of testis

terminal ring

flagellum

cells arranged in concentric layers

spermatozoon

4. Female Reproductive Elements

SCHEMATIC SECTION OF AN OVARY

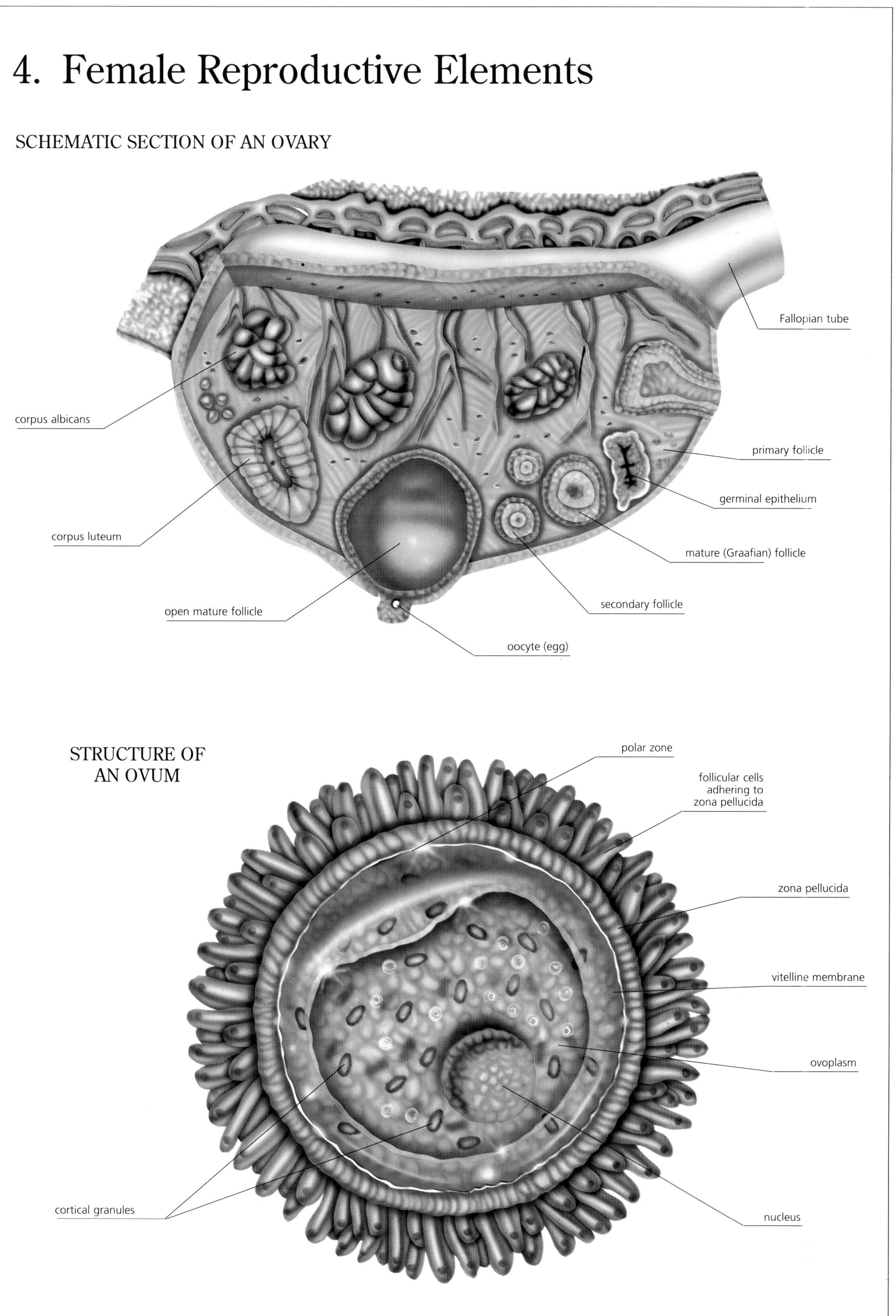

corpus albicans

Fallopian tube

corpus luteum

primary follicle

germinal epithelium

mature (Graafian) follicle

open mature follicle

secondary follicle

oocyte (egg)

STRUCTURE OF
AN OVUM

polar zone

follicular cells
adhering to
zona pellucida

zona pellucida

vitelline membrane

ovoplasm

cortical granules

nucleus

5. Birth and Pathway of an Unfertilized Ovum

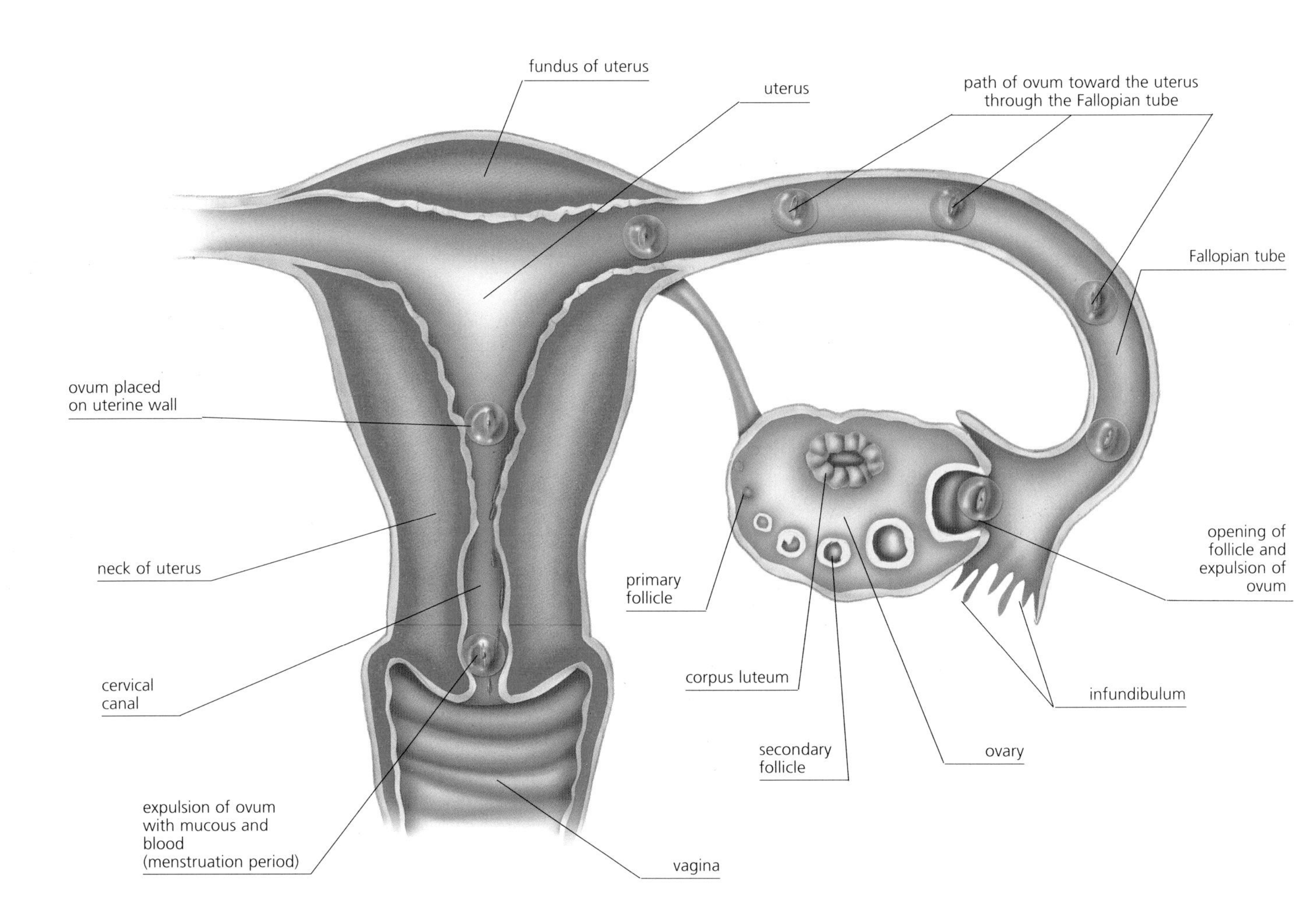

STAGES OF OVUM UNTIL ITS EXPULSION IN MENSTRUATION

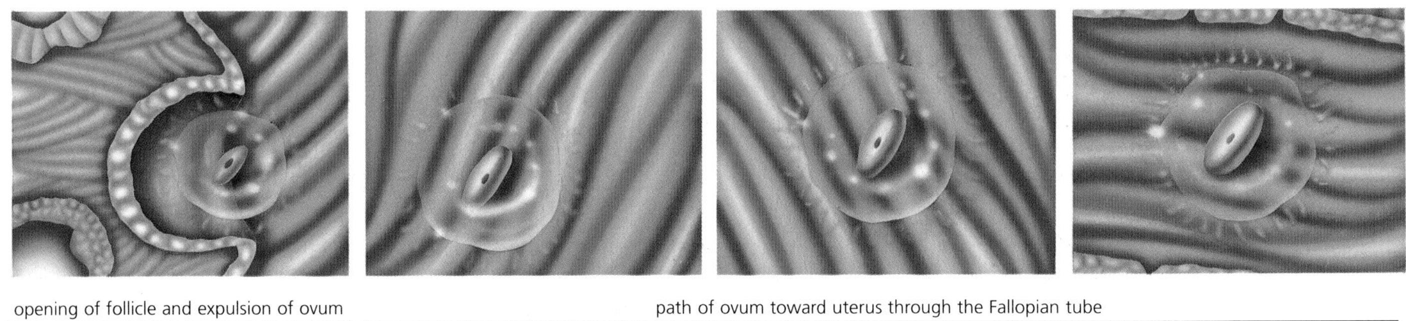

opening of follicle and expulsion of ovum

path of ovum toward uterus through the Fallopian tube

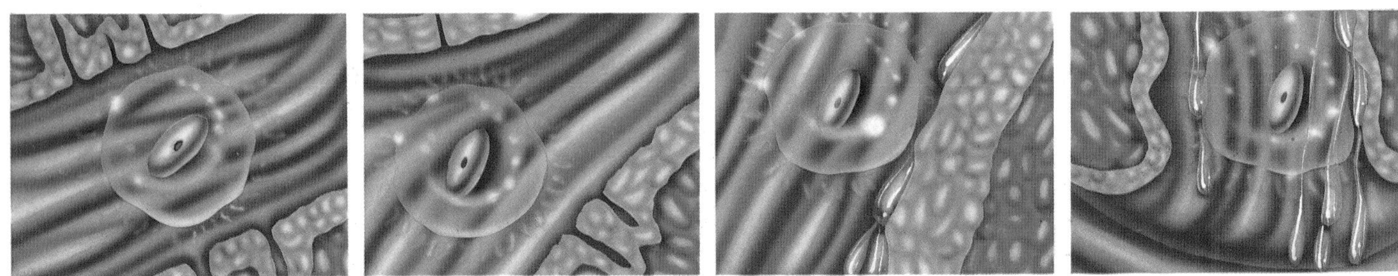
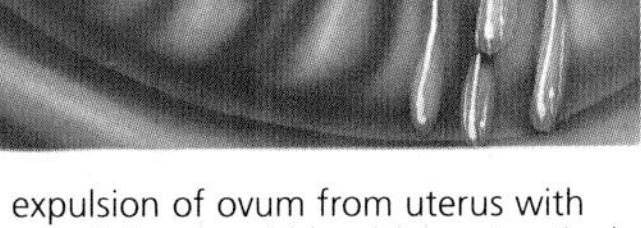

entrance of ovum into uterine cavity

deposition of ovum on uterine wall

expulsion of ovum from uterus with mucosal tissue and blood (menstruation)

6. Birth and Pathway of Fertilized Ovum

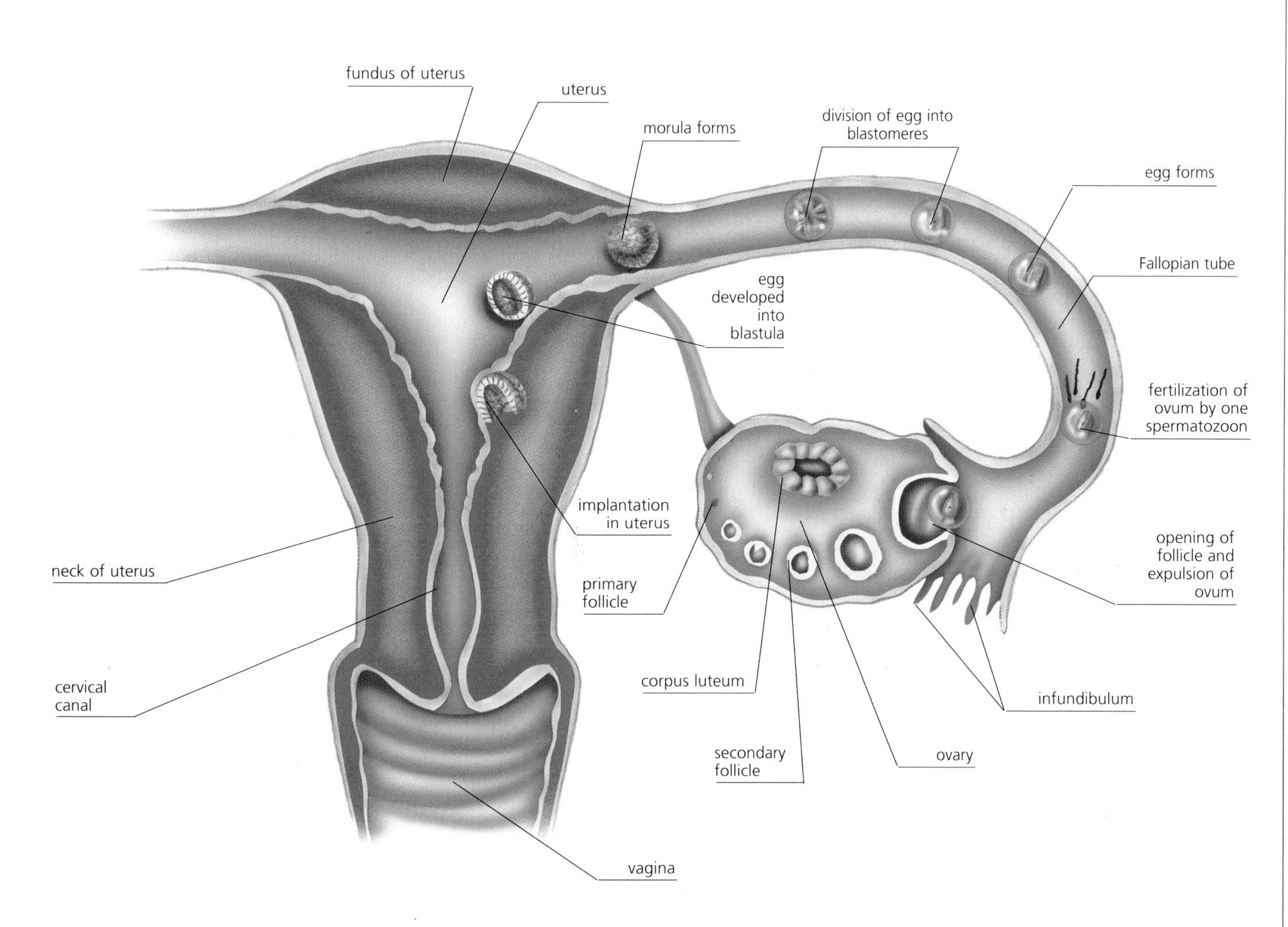

fundus of uterus

uterus

morula forms

division of egg into blastomeres

egg forms

egg developed into blastula

Fallopian tube

fertilization of ovum by one spermatozoon

neck of uterus

implantation in uterus

primary follicle

opening of follicle and expulsion of ovum

cervical canal

corpus luteum

secondary follicle

ovary

infundibulum

vagina

STAGES OF THE GENETIC MATERIAL (OVUM + SPERMATOZOON) UNTIL ITS IMPLANTATION IN THE UTERUS

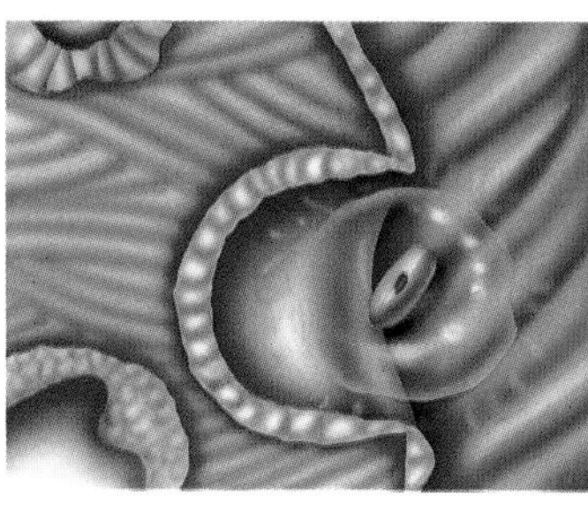
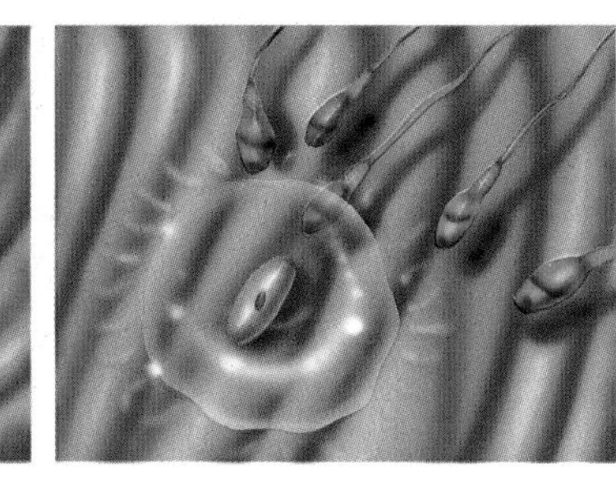
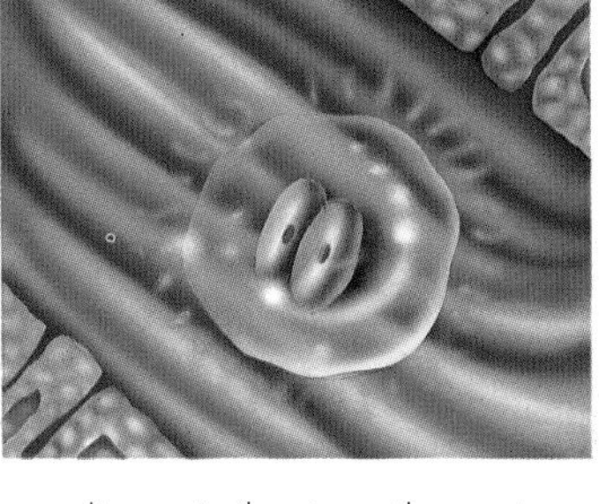
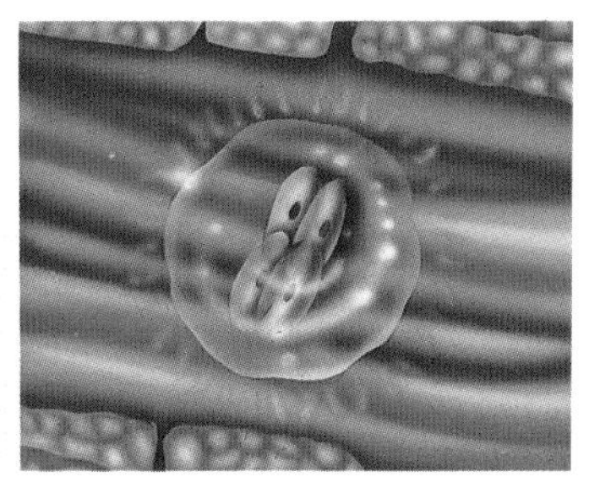

opening of follicle and expulsion of ovum

fertilization of ovum by one spermatozoon formation of zygote

on its way to the uterus, the zygote divides into two cells (blastomeres)

on its way to the uterus, the zygote divides into four cells

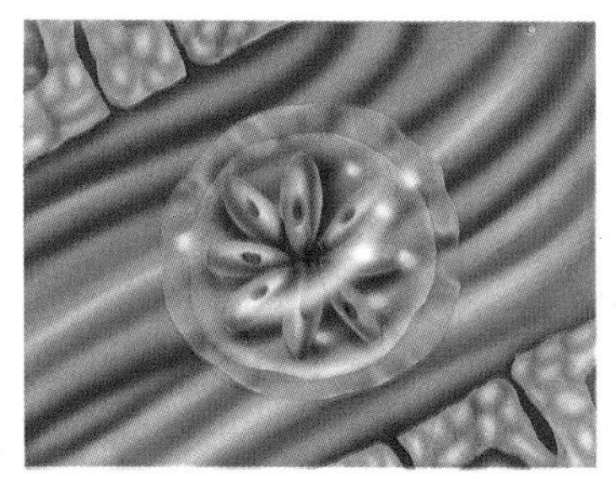
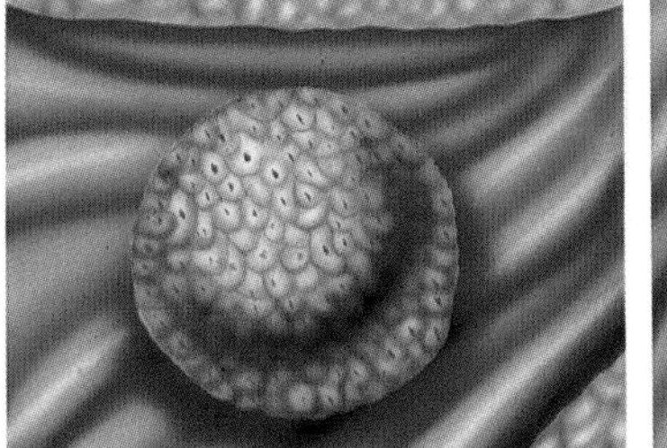
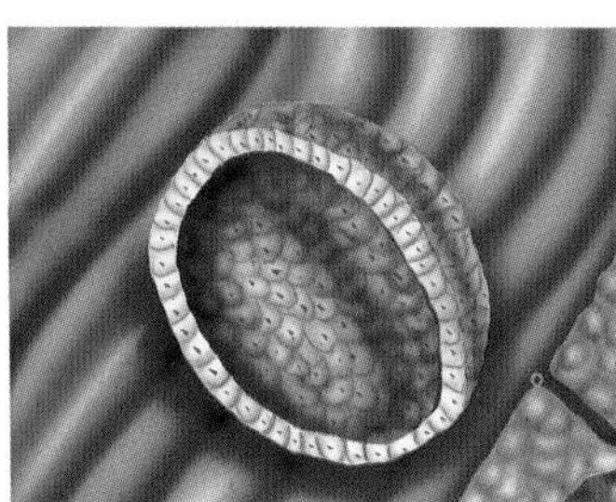
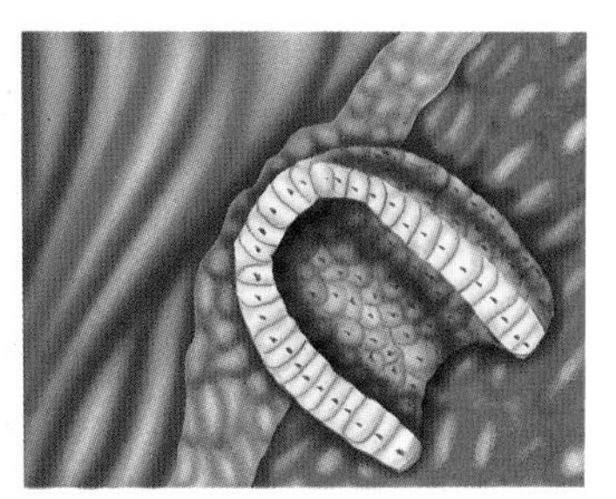

on its way to the uterus, the zygote divides into eight cells

formation of morula, a group of 32 cells

morula with blastomeres in blastula stage

nesting of morula on uterine wall

7. Menstrual Cycle

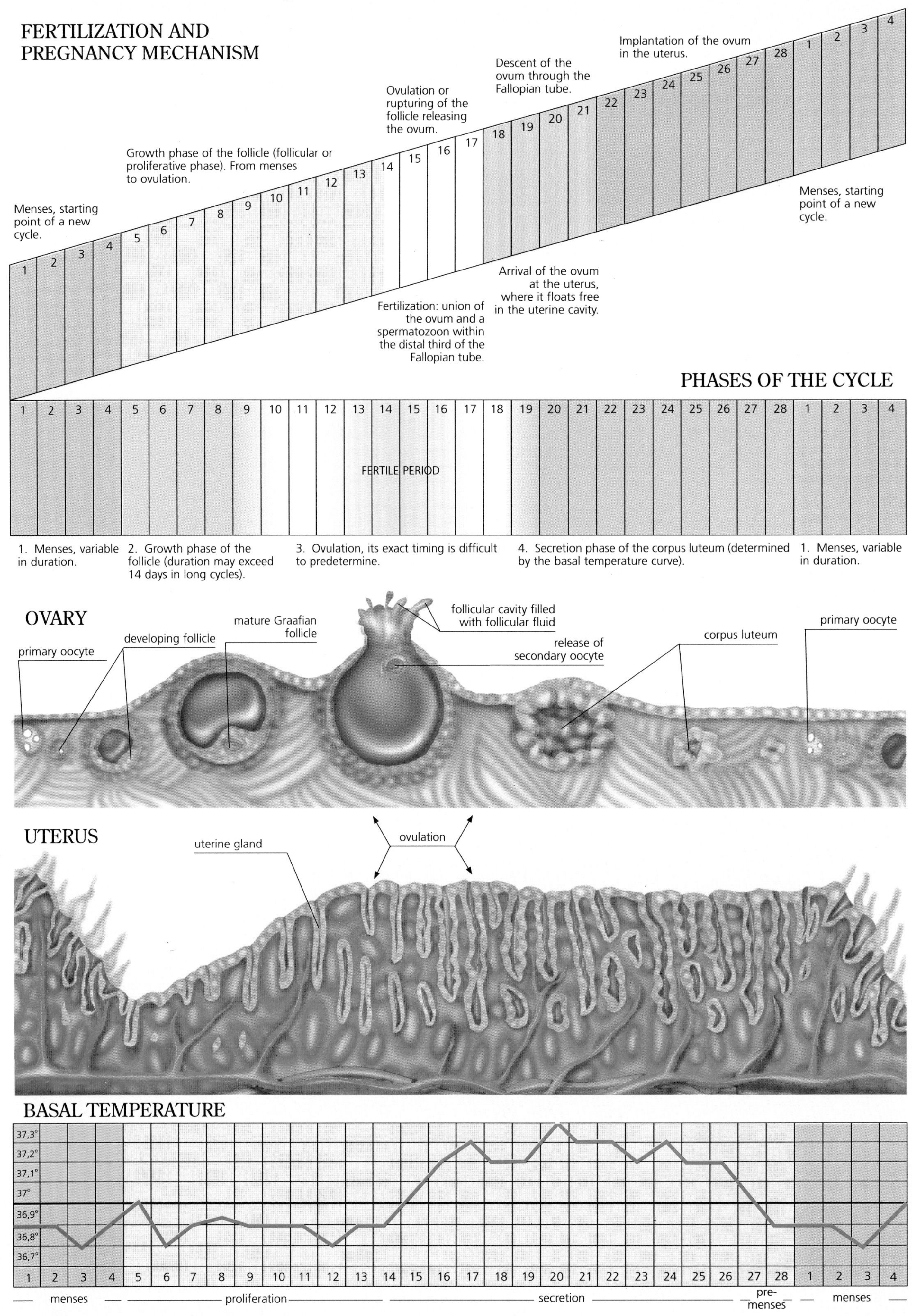

FERTILIZATION AND PREGNANCY MECHANISM

Menses, starting point of a new cycle.

Growth phase of the follicle (follicular or proliferative phase). From menses to ovulation.

Ovulation or rupturing of the follicle releasing the ovum.

Fertilization: union of the ovum and a spermatozoon within the distal third of the Fallopian tube.

Arrival of the ovum at the uterus, where it floats free in the uterine cavity.

Descent of the ovum through the Fallopian tube.

Implantation of the ovum in the uterus.

Menses, starting point of a new cycle.

PHASES OF THE CYCLE

FERTILE PERIOD

1. Menses, variable in duration.
2. Growth phase of the follicle (duration may exceed 14 days in long cycles).
3. Ovulation, its exact timing is difficult to predetermine.
4. Secretion phase of the corpus luteum (determined by the basal temperature curve).
1. Menses, variable in duration.

OVARY

primary oocyte
developing follicle
mature Graafian follicle
follicular cavity filled with follicular fluid
release of secondary oocyte
corpus luteum
primary oocyte

UTERUS

uterine gland
ovulation

BASAL TEMPERATURE

	37,3°
	37,2°
	37,1°
	37°
	36,9°
	36,8°
	36,7°

menses — proliferation — secretion — pre-menses — menses

84

8. Fertilization

FERTILIZATION AND NESTING

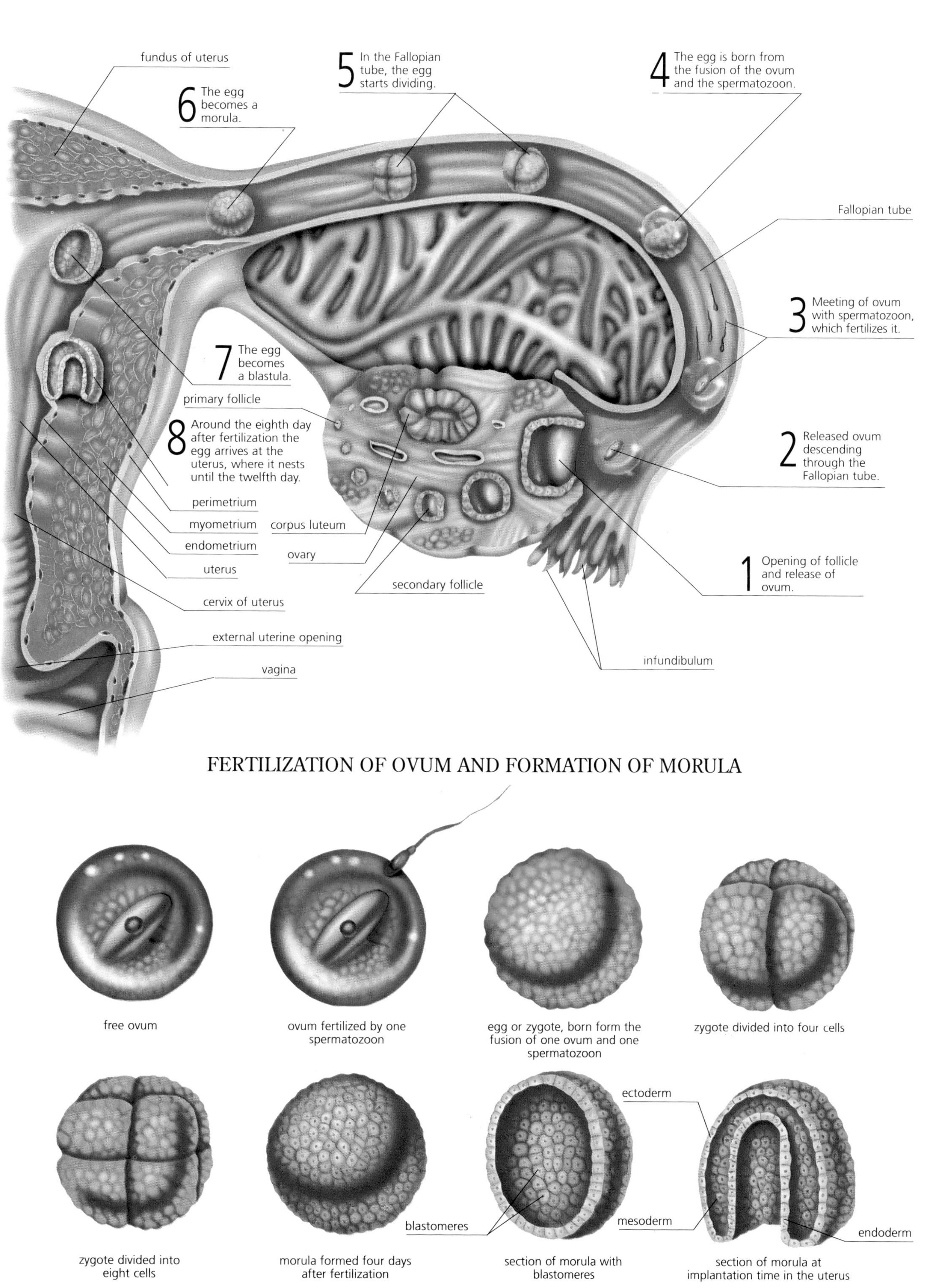

fundus of uterus

5 In the Fallopian tube, the egg starts dividing.

4 The egg is born from the fusion of the ovum and the spermatozoon.

6 The egg becomes a morula.

Fallopian tube

7 The egg becomes a blastula.

3 Meeting of ovum with spermatozoon, which fertilizes it.

primary follicle

8 Around the eighth day after fertilization the egg arrives at the uterus, where it nests until the twelfth day.

perimetrium

myometrium   corpus luteum

endometrium   ovary

2 Released ovum descending through the Fallopian tube.

uterus

cervix of uterus

secondary follicle

1 Opening of follicle and release of ovum.

external uterine opening

vagina

infundibulum

FERTILIZATION OF OVUM AND FORMATION OF MORULA

free ovum

ovum fertilized by one spermatozoon

egg or zygote, born form the fusion of one ovum and one spermatozoon

zygote divided into four cells

zygote divided into eight cells

morula formed four days after fertilization

section of morula with blastomeres

blastomeres

ectoderm

mesoderm

endoderm

section of morula at implantation time in the uterus

9. Placenta and Development of Fetus

SECTION OF PLACENTA

umbilical cord

umbilical vein

umbilical artery

chorionic villi

chorionic plate

chorion

pond of maternal blood

trophoblast

septum

decidua basalis

maternal blood vessel

myometrium

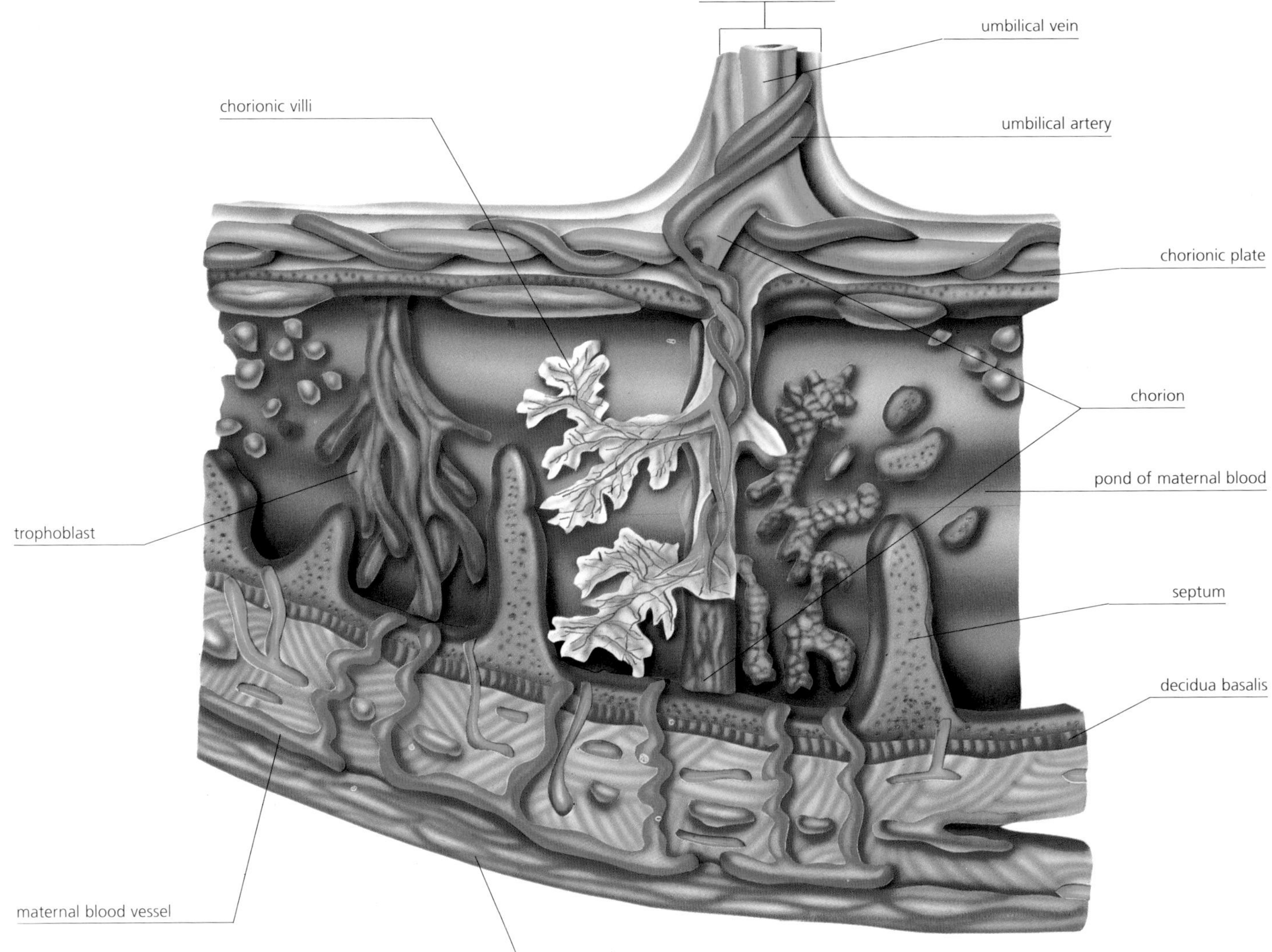

DEVELOPMENT OF FETUS INSIDE MATERNAL WOMB

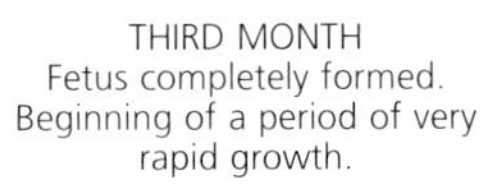

THIRD MONTH
Fetus completely formed.
Beginning of a period of very
rapid growth.

FIFTH MONTH
Fetus moves actively and
reacts to sound.

SEVENTH MONTH
Important maturation of
internal organs. Fetus capable
of survival.

NINTH MONTH
Fetus is completely developed
and positioned within
maternal pelvis for birth.

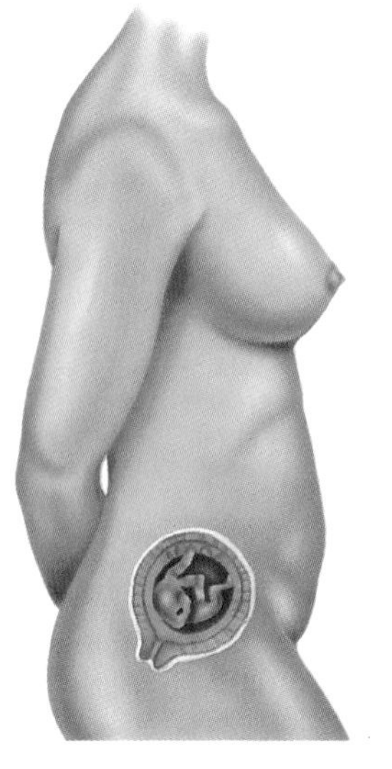

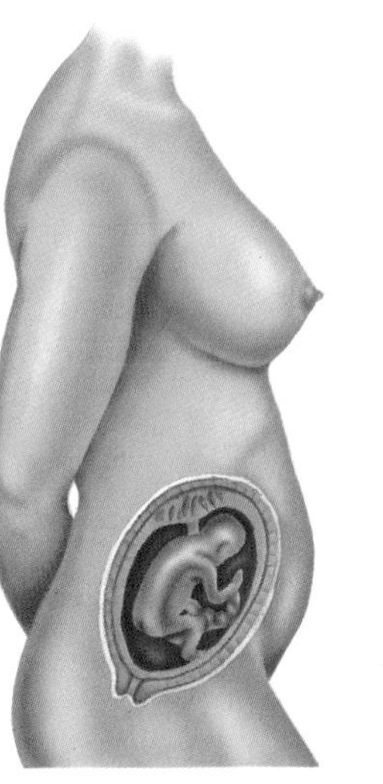

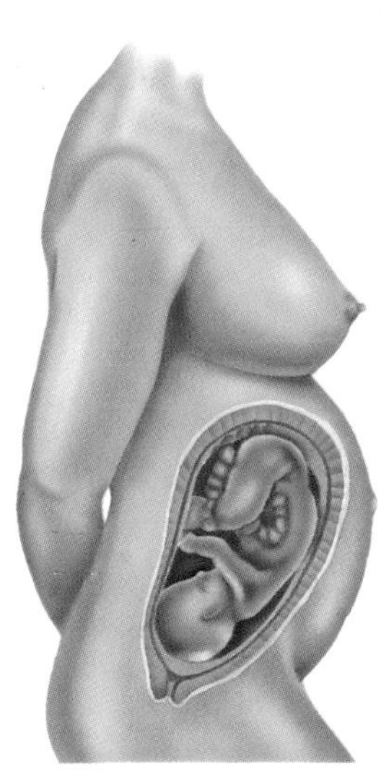

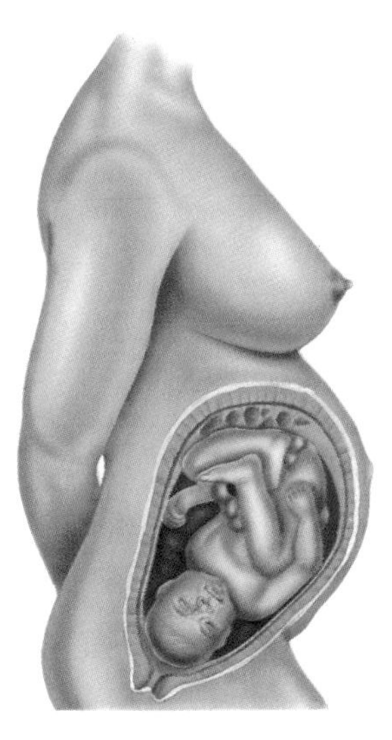

10. The Fetus During Gestation

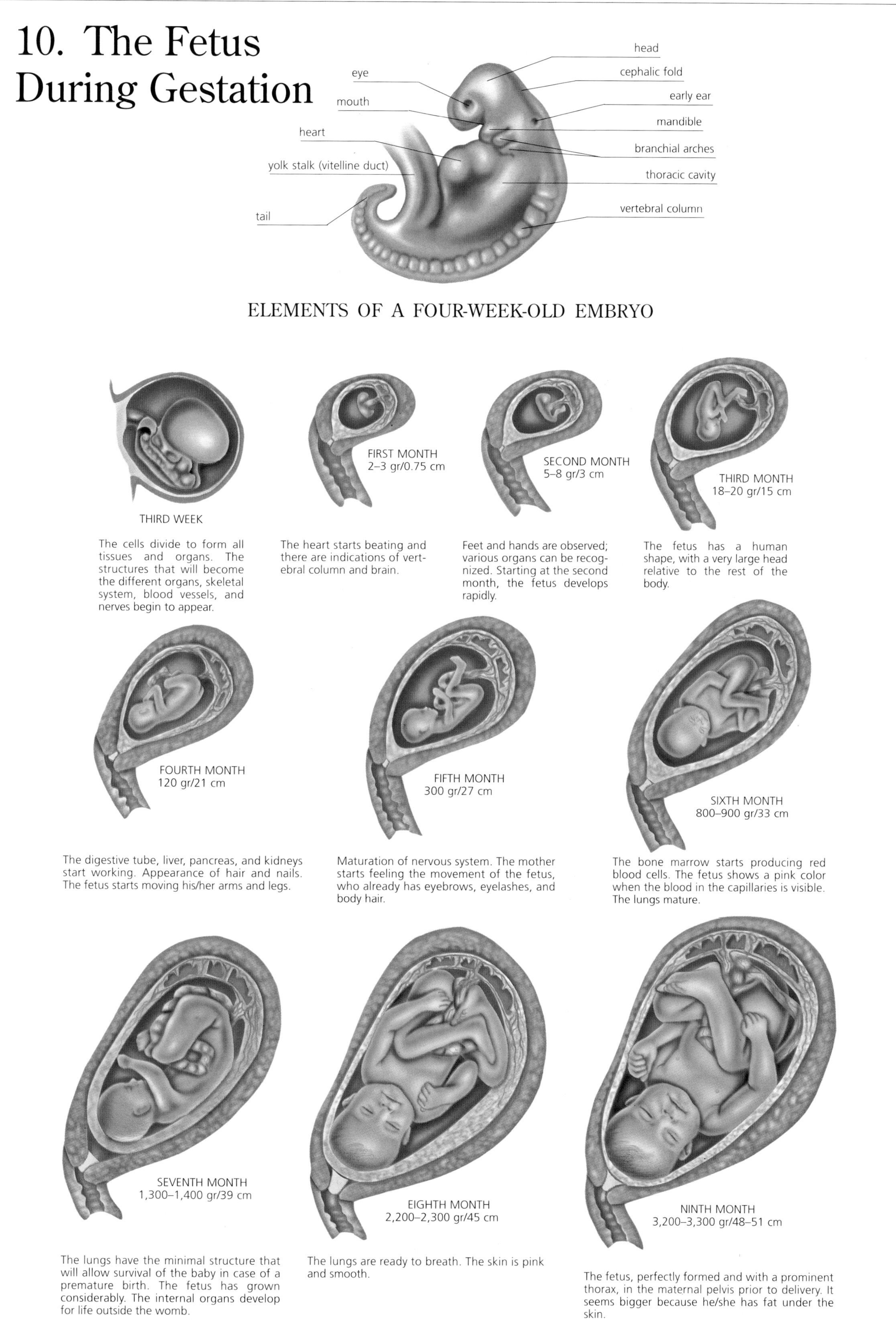

eye
mouth
heart
yolk stalk (vitelline duct)
tail
head
cephalic fold
early ear
mandible
branchial arches
thoracic cavity
vertebral column

ELEMENTS OF A FOUR-WEEK-OLD EMBRYO

THIRD WEEK

The cells divide to form all tissues and organs. The structures that will become the different organs, skeletal system, blood vessels, and nerves begin to appear.

FIRST MONTH
2–3 gr/0.75 cm

The heart starts beating and there are indications of vertebral column and brain.

SECOND MONTH
5–8 gr/3 cm

Feet and hands are observed; various organs can be recognized. Starting at the second month, the fetus develops rapidly.

THIRD MONTH
18–20 gr/15 cm

The fetus has a human shape, with a very large head relative to the rest of the body.

FOURTH MONTH
120 gr/21 cm

The digestive tube, liver, pancreas, and kidneys start working. Appearance of hair and nails. The fetus starts moving his/her arms and legs.

FIFTH MONTH
300 gr/27 cm

Maturation of nervous system. The mother starts feeling the movement of the fetus, who already has eyebrows, eyelashes, and body hair.

SIXTH MONTH
800–900 gr/33 cm

The bone marrow starts producing red blood cells. The fetus shows a pink color when the blood in the capillaries is visible. The lungs mature.

SEVENTH MONTH
1,300–1,400 gr/39 cm

The lungs have the minimal structure that will allow survival of the baby in case of a premature birth. The fetus has grown considerably. The internal organs develop for life outside the womb.

EIGHTH MONTH
2,200–2,300 gr/45 cm

The lungs are ready to breath. The skin is pink and smooth.

NINTH MONTH
3,200–3,300 gr/48–51 cm

The fetus, perfectly formed and with a prominent thorax, in the maternal pelvis prior to delivery. It seems bigger because he/she has fat under the skin.

11. Abdomen of a Pregnant Woman

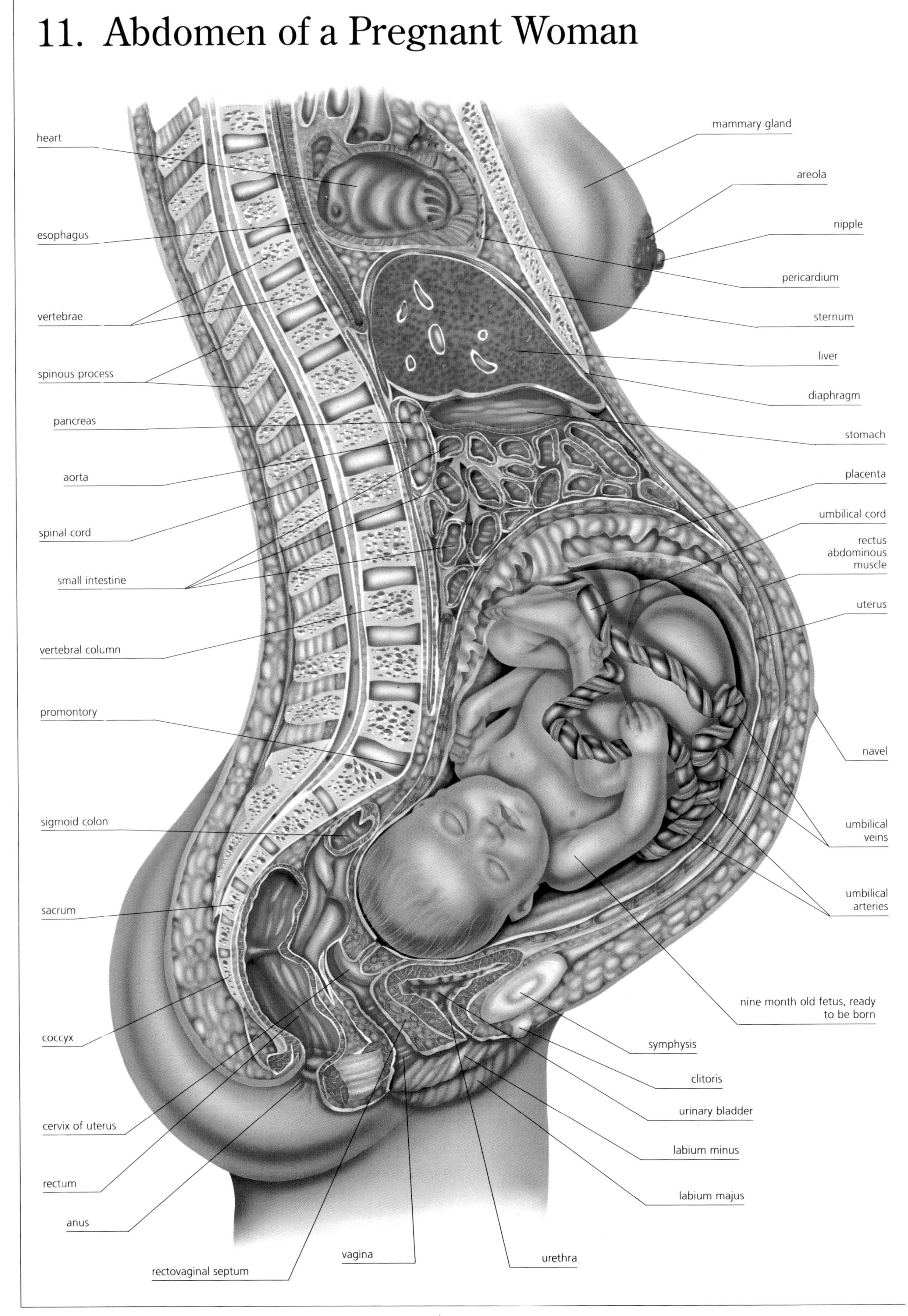

heart

esophagus

vertebrae

spinous process

pancreas

aorta

spinal cord

small intestine

vertebral column

promontory

sigmoid colon

sacrum

coccyx

cervix of uterus

rectum

anus

rectovaginal septum

vagina

urethra

mammary gland

areola

nipple

pericardium

sternum

liver

diaphragm

stomach

placenta

umbilical cord

rectus abdominous muscle

uterus

navel

umbilical veins

umbilical arteries

nine month old fetus, ready to be born

symphysis

clitoris

urinary bladder

labium minus

labium majus

12. The Placenta During Birth

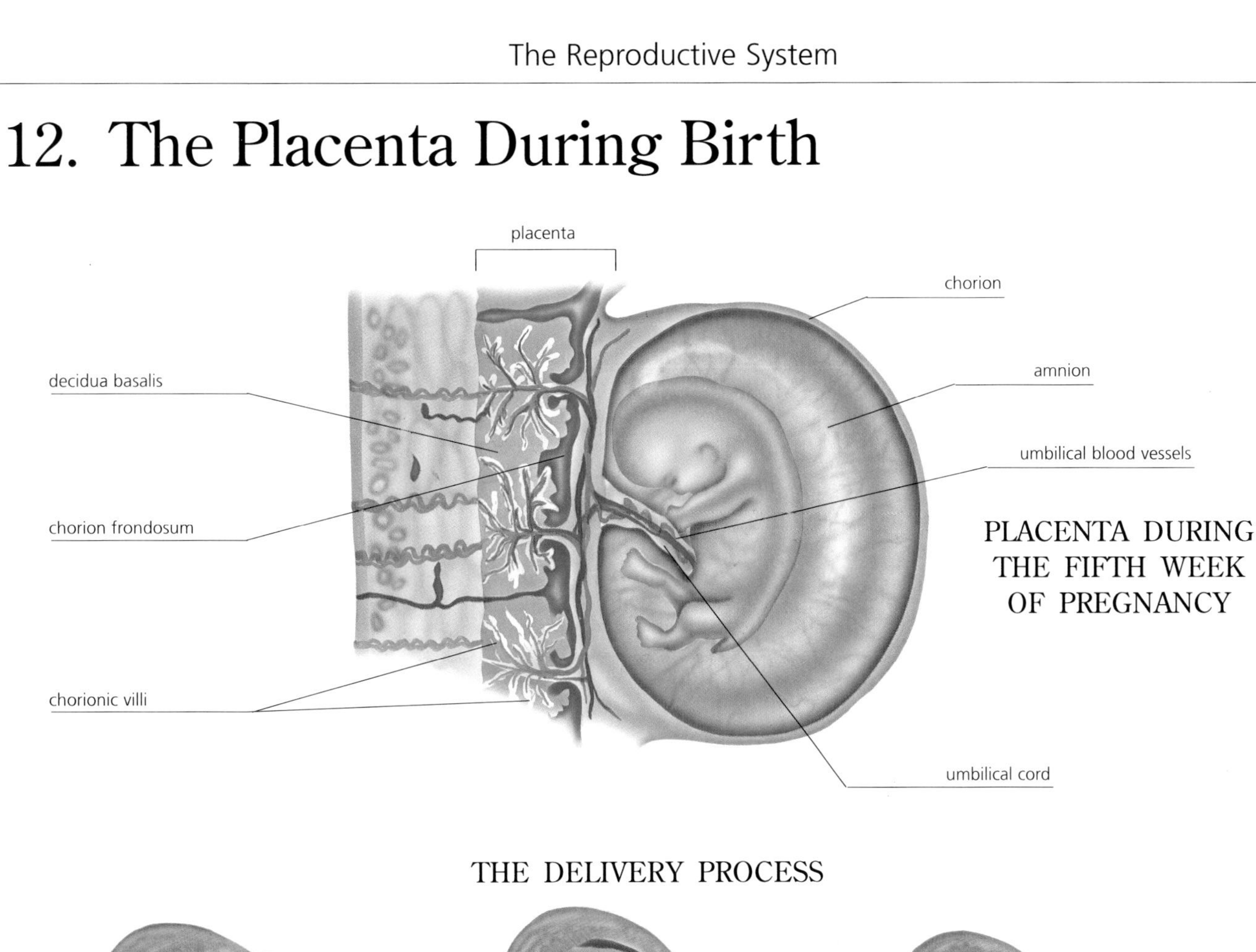

placenta

chorion

decidua basalis

amnion

chorion frondosum

umbilical blood vessels

PLACENTA DURING THE FIFTH WEEK OF PREGNANCY

chorionic villi

umbilical cord

THE DELIVERY PROCESS

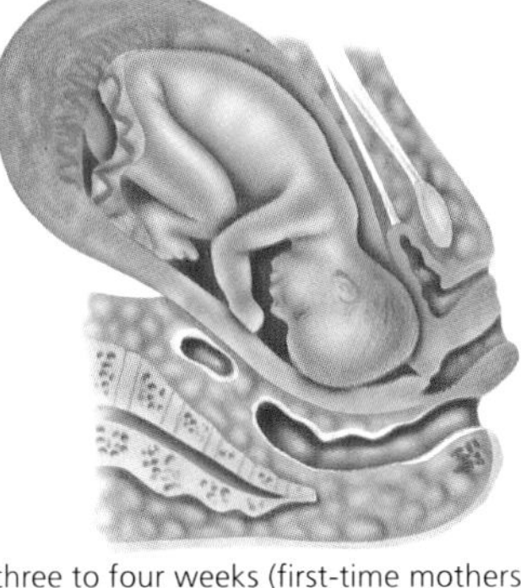
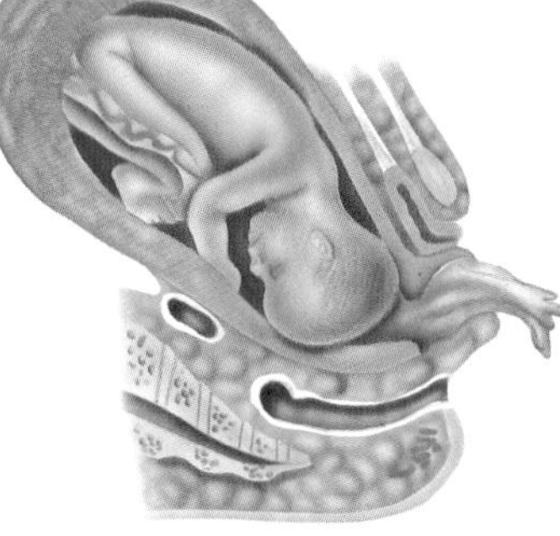
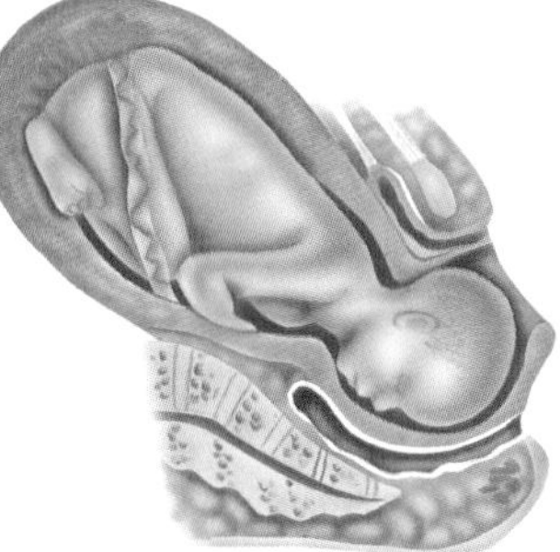

1 In three to four weeks (first-time mothers) or in a few hours (mother with previous births) before delivery, the fetus head positions at the opening of the pelvis.

2 The uterine muscles contract irregularly and with varying intensity. These contractions break the fetal membrane and the liquid inside (about two liters) pours out. The baby is pushed outside.

3 The cervix of the uterus dilates to about 10 cm to facilitate the exit of the fetus. Contractions are intermittent and growing in strength.

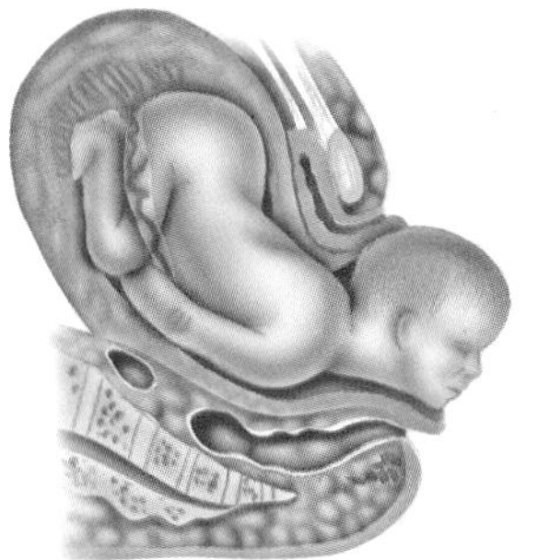
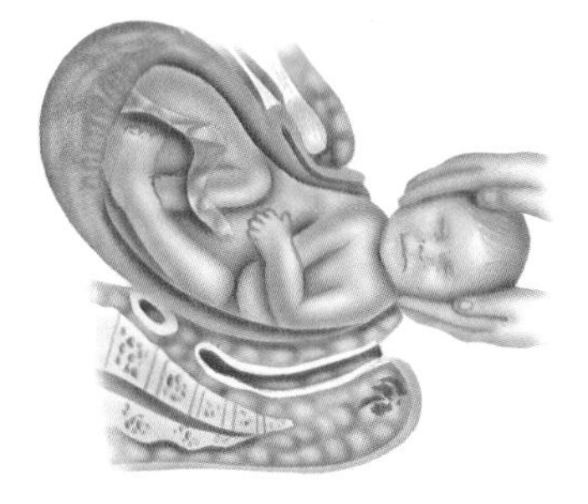
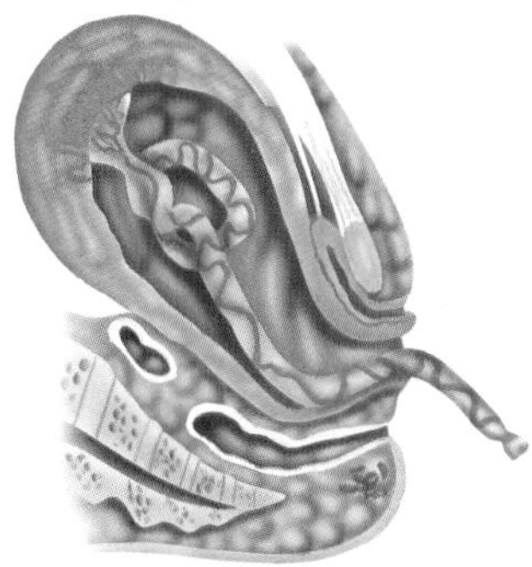

4 In a normal delivery, the head of the fetus exits first. If the mother is not fully dilated, it is necessary to incise the perineum to avoid complications.

5 After the head is outside, the baby's body turns and exits. the length of this phase, as well as all the others of this process, is variable.

6 Once the baby is outside the mother's womb, he/she is still attached to the placenta through the umbilical cord, which has to be cut. The placenta along with any residual matter are inside the maternal womb.

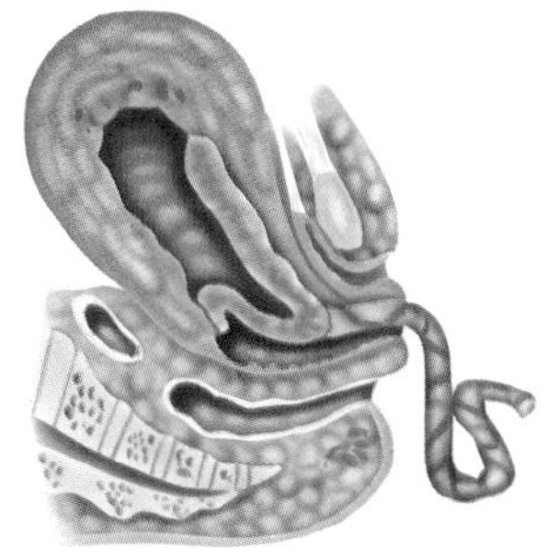

7 Strong contractions of the uterine muscles expel the umbilical cord, the placenta, and any residual matter to the exterior approximately 15 minutes after birth.

8 The placenta and umbilical cord are expelled from the maternal womb after birth. Labor is finished.

13. Breast and Mammary Glands

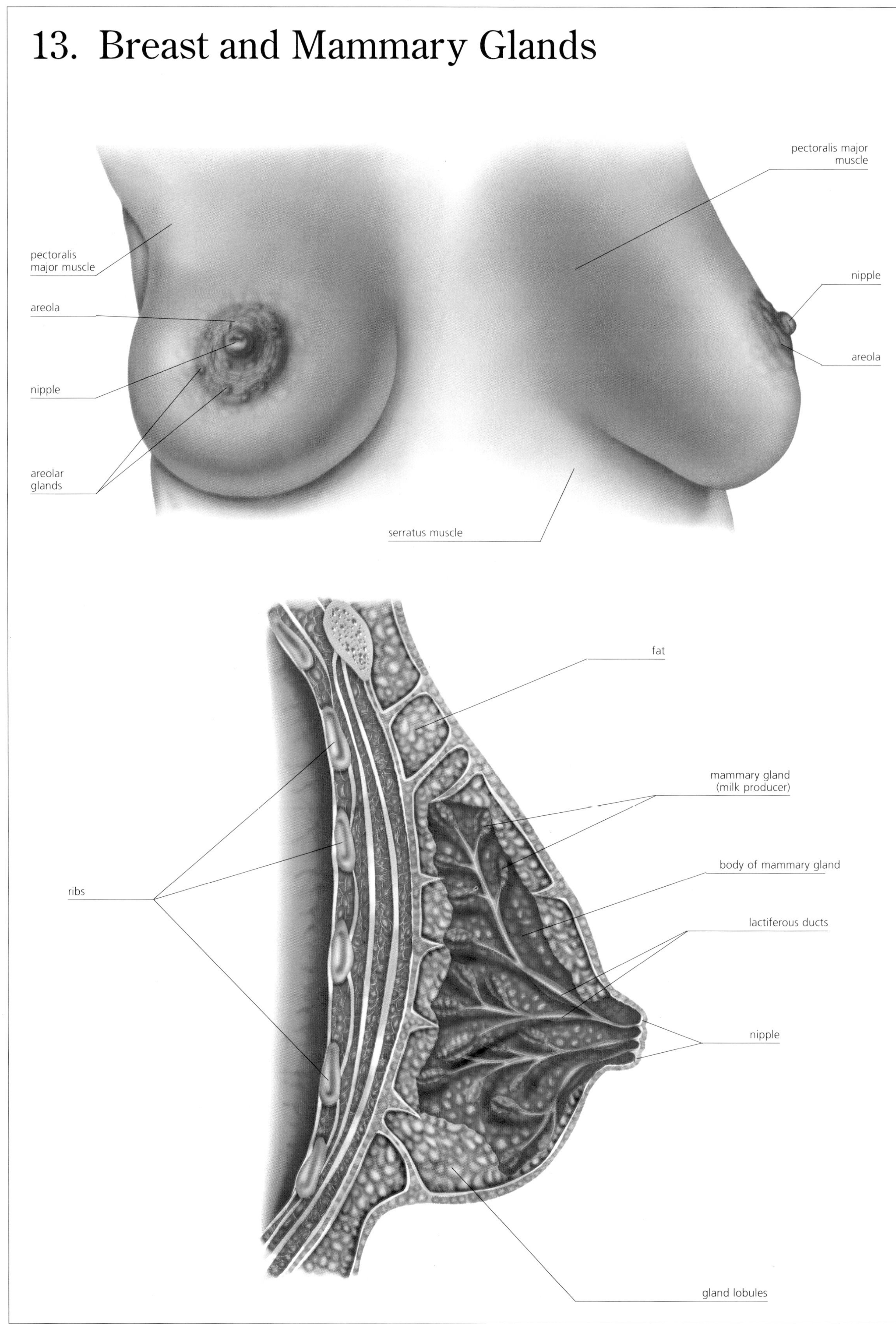

pectoralis major muscle

pectoralis major muscle

nipple

areola

areola

nipple

areolar glands

serratus muscle

fat

mammary gland (milk producer)

body of mammary gland

ribs

lactiferous ducts

nipple

gland lobules

Index

A

abdomen 6, 7, 88
abdomen of a pregnant woman 88
abdomen, aponeurosis of external oblique 26
abdomen, external abdominal oblique muscle 24
abdomen, sheath of rectus abdominis muscle of 26
abducens nerve 54
abductor 24
abductor digiti minimi muscle 24, 28
abductor magnus 24
abductor minimus 24
abductor pollicis longus muscle 24, 25
abductor pollicis muscle 28
absorption of fat 31
accessory nerve 54
acetabulum 20
Achilles tendon 28
aclitus of larynx 68
aconeus muscle 25, 26
acromion 14, 15, 16, 19, 26
acrosome 80
adductor magnus muscle 25
adductor pollicis muscle 28
adenine 11
alveolar air 46
alveolar artery 32
alveolar bone 41, 42
alveolar epithelium 46
alveolar vein 42
alveolus, capillary net of 46
alveolus, connective tissue of 46
alveolus, elastic fibers of 46
alveolus, gas exchange in 46
alveolus, process of blood oxygenation developed in pulmonary 46
amnion 89
ampulla of uterine tube 79
ampulla, anterior osseous of ear 65
ampulla, divert deferens of urinary bladder 77
ampulla, lateral osseous of ear 65
ampulla, posterior osseous of ear 65
ampulla, rectum 78
anal region 9
anatomy of the man 6
anatomy of the woman 7
angle, mastoid 17
ankle 6
annulus fribrosus of myocardium, left 48
antebrachial region 9
anterior antebrachial region 8, 9
anterior chamber, angle 59
anterior intercondylar area of tibia 20
anterior interventricular branch of left anterior artery 49
anterior part of liver 36
anterior view of the elbow 8
anterior view of the knee 8
anterior view of the leg 8, 9
anterior view of the neck 8
anterolateral section of heart 48
antihelix, root of 64
anus 31, 78, 79, 88
aorta 50
aorta, ascending part of 48, 49
aorta, bifurcation of 47

aorta, descending part of 49
aorta, thoracic part of 35
apex 17
apex of bladder 77
apex of head of fibula 20
apex of heart 48, 49
apex of lung 43
apex of patella 20
apex of sacrum bone 22
apex of spinous process of the seventh cervical vertebra 23
apex of tongue 68
aponeurosis of a muscle 29
aponeurosis, bicipital 24
aponeurosis, dorsal 28
apophysis, orbital of palatine bone 61
appendix fibrosa of liver 37
appendix of epididymis 80
appendix of testis 74, 78, 80
appendix vesiculosa 74, 79
appendix, vermiform 31, 39
arachnoidea 59, 60
arch of aorta 47, 49
arch of atlas, ventral surface of posterior arch 23
arch of axis, vertebral lamina of 22
arch of axis, vertebral pedicle of 22
arch of cricoid cartilage 45
arch of fifth lumbar vertebra, vertebral pedicle of 23
arch of fifth thoracic vertebra, vertebral pedicle of 22
arch of fourth cervical vertebra, vertebral lamina of 23
arch of fourth cervical vertebra, vertebral pedicle of 23
arch of fourth lumbar vertebra, vertebral lamina of 23
arch of fourth lumbar vertebra, vertebral pedicle of 23
arch of seventh cervical vertebra, vertebral lamina of 23
arch of seventh cervical vertebra, vertebral pedicle of 23
arch, anterior arch of atlas 23
arch, branchial 87
arch, crural 24
arch, infraorbital 17
arch, palatoglossal 68
arch, plantar 6, 7
arch, posterior arch of atlas 23
arch, supraorbital 47
area of brain, striated 63
area of cortex, occipital lobe striated 62
area of perception of acid taste 69
area of perception of salty taste 69
area of perception of sour taste 69
area of perception of sweet taste 69
area of tunica mucosa of stomach wall 33
areas, body 8, 9
areola 88, 90
arm (brachium) 19
arm of fetus 87
arm, medial cutaneous nerve of 55
arrival of egg to the uterus 84
arteriole of a main artery 52
arteriole, afferent glomerular 76
artery 41, 47, 52, 71
artery of bronchiole (branch of pulmonary artery) 46
artery of caudate lobe 37
artery, adventitia of a main 52

artery, adventitia of an 52
artery, alveolar 42
artery, anterior branch of renal 75
artery, anterior ciliary of eye 59
artery, anterior interventricular branch of left coronary 49
artery, anterior tibial 47
artery, arteriole of main 52
artery, atlantoccipital 64
artery, bifurcation of pulmonary trunk 49
artery, brachial 47
artery, central of retina 59, 60
artery, circumflex branches of left coronary 49
artery, collagen fiber of a main 52
artery, common carotid 47, 49
artery, common hepatic 35
artery, common iliac 47
artery, coronary 47
artery, deep artery of penis 77, 78
artery, digital 47
artery, dorsal artery of foot 47
artery, dorsal artery of penis 78
artery, elastic fiber of a main 52
artery, elastic lamina of an 52
artery, endothelium of an 52
artery, external elastic lamina of a main 52
artery, external iliac 47
artery, external layer of an 52
artery, femoral 47
artery, fibular 47
artery, gastric 47
artery, gastroduodenal 35
artery, hepatic 31, 47
artery, ileal 35
artery, inferior mental 41
artery, inferior mesenteric 35, 47
artery, internal carotid 62
artery, internal elastic lamina of an 52
artery, internal iliac 47
artery, internal layer of an 52
artery, jejunal 35
artery, lateral plantar 47
artery, left coronary 49
artery, left gastric 35
artery, left pulmonary 49
artery, lienal sulcus of 35, 40
artery, long posterior ciliary 59
artery, main of the circulatory system 47
artery, metatarsal 47
artery, middle colic 35
artery, middle layer of an 52
artery, muscular fiber of an 52
artery, muscular layer of an 52
artery, net of nerves of an 52
artery, ophthalmic 61, 62
artery, ovarian 35, 79
artery, popliteal 47
artery, posterior ciliary 60, 61
artery, posterior interventricular branches of right coronary 49
artery, posterior ramification of posterior renal 75
artery, posterior tibial 47
artery, proper hepatic 35, 37
artery, pulmonary 46, 47, 50
artery, radial 47
artery, renal 35, 75
artery, right coronary 48, 49
artery, right inferior suprarenal 75
artery, right middle suprarenal 75

artery, right pulmonary 49
artery, schematic section of an 52
artery, short posterior ciliary 59
artery, small arteries of an 52
artery, soft tissue of an 52
artery, splenic 35, 40, 47
artery, subclavian 47, 49
artery, sulcus of splenic 35
artery, superior mesenteric 35, 47
artery, superior suprarenal 75
artery, testicular 35, 78
artery, tunica intima of a main 52
artery, ulnar 47
artery, umbilical 86, 88
artery, valves of pulmonary 48
artery, vertebral 23
atlas, anterior articular facet of 23
atlas, anterior crus of 23
atlas, anterior tubercle of 23
atlas, articular facet of odontoid process of 23
atlas, foramen of transverse process of 23
atlas, fossa of axis for the transverse ligament of 23
atlas, inferior articular fossa of 23
atlas, lateral mass of 23
atlas, margin of glenoid cavities of 23
atlas, posterior crus of 23
atlas, posterior tubercle of 23
atlas, sulcus of vertebral artery of 23
atlas, superior articular fossa of 23
atlas, superior view of 23
atlas, transverse process of 23
atlas, ventral view of 23
atrium, left 48, 49, 50
atrium, right 48, 49, 50
auricular, greater nerve 58
auricularis anterior muscle 24
auricularis posterior muscle 25, 26, 27
auricularis superior muscle 24, 25
axilla (armpit) 6, 7, 72
axillary artery 47
axillary nerve 55, 58
axillary region 9
axillary vein 47
axis 15, 16, 21
axis, anterior articular facet of 22
axis, dorsal view of 22
axis, foramen of transverse process of 22
axis, fossa for transverse ligament of atlas 22
axis, inferior articular facet of 22
axis, inferior articular process of 22
axis, inferior articular surface of the vertebral body of 22
axis, junction of spinous process of 22
axis, lamina of vertebral arch of 22
axis, lateral view of 22
axis, odontoid process (dens) of 22
axis, pedicle of vertebral arch of 22
axis, spinous process of 22
axis, superior articular facet of 22
axis, transverse process of 22
axis, tuberosity of atlas for the insertion of transverse ligaments of 22
axis, vertebral body of 22
axis, vertebral foramen of 22
axon of a bipolar neuron 56
axon of a motor neuron 56
axon of a multipolar neuron 56

axon of a unipolar neuron 56
axon terminal 29

B

baby 89
back 6, 7
back of the hand 9
bacteria 11
Bartolino, gland of 74
base of cochlea 65
base of dorsal horn of spinal cord 73
base of lung 43
base of metacarpal bone 18
base of metatarsal bone 18
base of patella 20
base of phalanx of fingers 18
base of phalanx of toes 18
base of pyramid of Malpighi 75
base of sacrum 22
base of stapes 66
basilic vein 47
basophil 51
beat, phases of heart 50
bed of retina, capillary 59
bed, alveolar capillary 46
bed, ungueal 73
belly of epicranius muscle, frontal 27
belly of epicranius muscle, occipital 27
belly of omohyoid muscle, inferior 26
biceps brachii in resting phase 30
biceps brachii muscle 24, 25
biceps brachii tendon 24
biceps brachii, contracted 30
biceps brachii, relaxed 30
biceps femoris muscle 25
biceps femoris muscle, short head of 26
bicipital aponeurosis 24, 25
bifurcation of aorta 47
bifurcation of pulmonary trunk 49
bifurcation of trachea 43, 44, 45
bile 31
bipenniform muscle 29
bipenniform muscle, intermediate tendon of 29
birth of fertilized oocyte 83
bladder, apex of 77
bladder, bottom of 77
bladder, dorsal view of 77
bladder, male 78
bladder, mucosal folds of 77
bladder, muscular layer of body of 77
bladder, trigone of 77, 78
bladder, tunica mucosa of 77
bladder, urinary 74, 77, 78, 79, 88
bladder, uvula of 77, 78
bladder, ventral view of 77
blastomere 85
blastomeres, division of egg into 83
blastomeres, division of zygote into eight 83
blastomeres, division of zygote into four 83
blastomeres, division of zygote into two 83
blastomeres in morula, in blastula stage 83
blastula 85
blastula, egg transformed into 83
blastula, morula with blastomeres in 83
blood 51, 52, 82

blood cell, microscopic view of a 51
blood cell, red 51
blood cell, white 51
blood of fetus 87
blood, deoxygenated 46, 50
blood, maternal 86
blood, oxygenated 46, 50
blood, process of oxygenation developed in pulmonary alveolus, 46
blood, venous 46
blue-green algae 11
body area 8, 9
body of a bipolar neuron 56
body of a finger 73
body of a motor neuron 56
body of a multipolar neuron 56
body of a unipolar neuron 56
body of axis 22
body of brain, lateral geniculate 62, 63
body of brain, medial geniculate 63
body of clitoris 79
body of epididymis 80
body of gallbladder 37
body of ilium 20
body of incus 66
body of ischium 20
body of mammary gland 90
body of mandible 17
body of pancreas 34, 35
body of penis 80
body of penis, bulb of corpus spongiosum 78
body of penis, corpus spongiosum 74, 77, 78
body of pubis 20
body of retina, vitreous 59
body of stomach 32, 33
body of the fifth lumbar verterbra 23
body of the fifth thoracic vertebra 22
body of the fourth cervical verterbra 23
body of the fourth lumbar verterbra 23
body of the seventh cervical verterbra 23
body of the sixth thoracic verterbra 22
body of tongue 68
body of urinary bladder, muscular layer of 77
body of vertebrae 73
body, body areas of the human 9
body, ciliary 62
body, different types of skin of the human 71
body, human 69
body, inferior articular facet of vertebral 22, 23
body, intervertebral facet of vertebral 23
body, mamillary 63
body, mamillary, of brain 54
body, muscles of the human 24, 25, 26
body, muscular fiber of the human 29
body, orbital adipose 61
body, posterior view of ciliary 62
body, vitreous 59, 60
bone, a triple root tooth 41
bone, alveolar 41, 42

91

bone, anterior crus of stapes 66
bone, anterior foramina of sacrum 22
bone, anterior process of malleus 66
bone, anterior view of 22
bone, apex of sacrum 22
bone, auricular surface of sacrum 22
bone, backbone 13
bone, base of sacrum 22
bone, base of stapes 66
bone, body of incus 66
bone, calcaneus 16, 18
bone, capitate 14, 18, 28
bone, carpal 13, 14, 15, 16
bone, coccyx 13, 15, 16, 21, 22, 58
bone, cuboid 16, 18
bone, ethmoid 12, 16, 17, 67
bone, femur 16
bone, fibula 16, 28
bone, finger 73
bone, fingers 13
bone, foot 13, 18
bone, foramen of sacrum 22
bone, frontal 12, 14, 16, 17, 61, 67
bone, hand 13, 18
bone, head 12
bone, head of malleus 66
bone, hip (os coxae) 13, 20
bone, hook of triquetral 28
bone, humerus 16
bone, hyoid 16, 68
bone, ilium 13, 14, 20
bone, imprint of roots of teeth on mandible 41, 42
bone, imprint of roots of teeth on maxilla 41, 42
bone, incus 65, 66
bone, intermediate cuneiform 16
bone, interparietal 17
bone, ischium 14
bone, lacrimal 12, 16, 17
bone, lateral cuneiform 16
bone, lateral portion of sacrum 22
bone, lateral process of malleus 66
bone, lateral view of nasal 67
bone, lateral view of sacrum 22
bone, limb 13
bone, long crus of incus 66
bone, lower limb 20
bone, lunate 14, 18
bone, malar 12, 14, 16, 17
bone, malleus 65, 66
bone, mandible 16, 17, 31, 41, 67
bone, manubrium of malleus 66
bone, maxilla 12, 42, 61
bone, medial cuneiform 16
bone, median crest of sacrum 22
bone, metacarpal 13, 14, 15, 16
bone, metatarsal 13, 18
bone, nasal 12, 14, 16, 17, 67
bone, navicular 16, 18
bone, neck of malleus 66
bone, occipital 12, 15, 16, 17, 64
bone, palatine 12, 67
bone, parietal 12, 14, 15, 16, 17
bone, pisiform 18, 28
bone, posterior crus of stapes 66
bone, promontory of sacrum bone 22
bone, pubis 14
bone, radius 16, 18, 28
bone, sacrum 13, 14, 15, 16, 21, 57, 58, 78, 79
bone, scaphoid 14, 18
bone, short crus of incus 66
bone, skull 17
bone, sphenoid 14, 17, 61, 67
bone, stapes 65, 66
bone, superior articular process of sacrum 22
bone, sutural 17
bone, talus (ankle bone) 16, 18
bone, tarsus 13
bone, temporal 12, 14, 15, 16, 64
bone, tibia 16, 28
bone, toes 13
bone, transverse lines of sacrum 22
bone, trapezoid 14, 18
bone, trapezium 14, 18
bone, triquetral 14, 18, 28
bone, tuberosity of sacrum 22
bone, ulnar 16, 18, 28
bone, upper limb 19
bone, vomer 12, 67
bone, wing of sacrum 22
bone, zygomatic or malar 12, 14, 16, 17
bottom of the eye 60
bottom of urinary bladder 77
botton, synaptic of motor neuron 56
botton-shaped papilla 70
Bowman gland of 67
Bowman's capsule of 76
Bowman's capsule, basement membrane of 59
Bowman's capsule, space of 76
brachial artery 47
brachial plexus 58
brachial posterior region 8, 9
brachial region 9
brachialis anterior muscle 24, 25
brachialis anticus muscle 26
brachialis muscle 26
brachiocephalic vein 47
brachiocephalic vein, left 49
brachiocephalic vein, right 49
brachioradialis muscle 24, 25
brachium of superior colliculus of brain 63
brain 53, 54, 66, 73
brain of fetus 87
brain showing visual paths, inferior view of 63
brain, angular gyrus of 53
brain, anterior median sulcus of 54
brain, anterior perforated substance of 54
brain, brachium of superior colliculus of 63
brain, central sulcus of 53
brain, choroid plexus of lateral ventricle of 63
brain, cingulate sulcus of 53
brain, floculus of 54
brain, frontal pole of 53, 54
brain, frontal superior gyrus of 53
brain, hypophysis of 54
brain, inferior frontal sulcus of 53
brain, inferior parietal gyrus of 53
brain, inferior parietal lobe of 53
brain, inferior temporal sulcus of 54
brain, inferior view of 54
brain, infundibulum of 54
brain, interparietal sulcus of 53
brain, interpeduncular fossa of 54
brain, lateral geniculate body of 63
brain, lateral occipitotemporal gyrus of 53
brain, lateral root of 63
brain, lateral sulcus of 53
brain, longitudinal sulcus of 53, 54
brain, mammilary body of 54
brain, middle frontal gyrus of 53
brain, middle geniculate body of 63
brain, middle root of 63
brain, occipital gyrus of 53
brain, occipital pole of 53, 54
brain, olive of 54
brain, orbital gyrus of 53
brain, orbital sulcus of 54
brain, parietooccipital sulcus of 53
brain, pons of 54
brain, postcentral gyrus of 53
brain, postcentral sulcus of 53
brain, precentral sulcus of 53
brain, primary fiber of 70
brain, pyramid of 54
brain, pyramidal decussation of 54
brain, secondary fiber of 70
brain, striated area of 63
brain, superior frontal sulcus of 53
brain, superior parietal lobule of 53
brain, superior temporal sulcus of 53
brain, superior view of 54
brain, supramarginal gyrus of 53
brain, temporal pole of 54
brain, tertiary fiber of 70
branch of intercostal nerve, anterior cutaneous 58
branch of intercostal nerve, lateral cutaneous 58
branch of ischium 20
branch of lateral cutaneous nerve of forearm 58
branch of left coronary artery, anterior interventricular 49
branch of left coronary artery, circumflex 49
branch of posterior cutaneous nerve of forearm 58
branch of pubis, inferior 20
branch of pulmonary artery 46
branch of pulmonary artery of bronchioles 46
branch of radial nerve, dorsal digital branches of 55
branch of radial nerve, superficial 55
branch of renal artery, anterior 75
branch of right coronary artery, posterior interventricular 49
branch of spinal cord, interganglionar 79
branch of superficial branch of radial nerve, dorsal digital 55
branch of superficial cervical plexus, mastoid 55
branch of superficial cervical plexus, supraclavicular 55, 58
branch of superficial cervical plexus, supracromial 55, 58
branch of thoracic nerves, posterior 58
branch of tibial nerve, plantar 58
branch, anterior cutaneous 58
branch, palmar 47
breast 7, 88
breathing, mechanism of 46
bregma 17
broad muscle 29
bronchial nerve 46
bronchiole 43
bronchiole, branch of pulmonary artery of 46
bronchiole, cartilage of 46
bronchiole, mucous gland of 46
bronchiole, terminal 46
bronchus 43
bronchus, anterior view of 45
bronchus, left primary 44, 45
bronchus, projection on wall of lungs 43
bronchus, right primary 44, 45
buccal region 9
buccinator muscle 24, 26
bud, villi of taste 69
bulb of corpus spongiosum of penis 78
bulb of jugular vein, 64
bulb of penis 77, 78
bulb of vestible of vagina 74
bulb, hair 71, 72
bulb, olfactory 54, 63, 67
bulbourethral gland, duct of 78
bursa ungueal 73
buttock 6, 7

C

calcaneal region 8, 9
calcaneous, peroneal of trochlea 18
calcaneus bone 15, 16, 18
calcaneus, tuberosity of 18, 25
calyx, major of kidney 75
calyx, minor of kidney 75
canal of an incisor tooth, root 41
canal of tensor tympanic muscle 66
canal, anterior semicircular of ear 66
canal, anterior semicircular of osseous labyrinth 65
canal, central, of spinal cord 73
canal, cervical 79, 82, 83
canal, infraorbital 42
canal, lateral semicircular of osseous labyrinth of ear 65
canal, mental 41
canal, posterior semicircular of osseous labyrinth of ear 65
canal, pyloric 34
canaliculus, inferior lacrimal 61
canaliculus, superior lacrimal 61
canine tooth 41, 42
capillary of fetus 87
capillary, arterial of a hair 72
capillary, blood 46
capillary, cellular layer of a blood 52
capillary, endothelial layer of a blood 52
capillary, glomerular 76
capillary, internal layer of a blood 52
capillary, schematic section of a blood 52
capillary, venous of a hair 72
capitate bone 14, 18, 28
capitulum of humerus 19
capsule of Bowman 76
capsule of Bowman, basement membrane of 76
capsule of kidney, adipose 75
capsule of kidney, fibrous 75
carbon dioxide 46
cardiac stomach 31, 32
cardiac vein, great 49
cardiac vein, middle 49
carina 43, 44, 45
carotid artery, common 47, 49
carotid artery, internal 62
carpus 18, 19
carpus, bones of 13, 16
carpus, extensor carpi radialis brevis muscle of 26
carpus, extensor carpi radialis longus muscle of 26
carpus, transverse ligament of 28
cartilage of trachea 43
cartilage, acoustic 64
cartilage, arch of cricoid 45
cartilage, arytenoid 45
cartilage, bronchial 45
cartilage, corniculate 45
cartilage, cricoid 43
cartilage, greater alar 67
cartilage, lamina of cricoid 45
cartilage, lamina of thyroid 45
cartilage, lateral crus of greater alar 67
cartilage, lesser alar 67
cartilage, nasal 67
cartilage, septal 67
cartilage, thyroid 43, 68
cartilage, tracheal 45
caruncula lacrimalis 61
cavity (cavum) of concha of external ear 64
cavity of atlas, margin of glenoid 23
cavity of incisor tooth 41
cavity of incisor tooth, central 41
cavity of premolar tooth, central 42
cavity of scapula, glenoid 19
cavity of tympanus, facial duct of 66
cavity of tympanus, mastoid antrum of 66
cavity of tympanus, mastoid wall of 66
cavity of tympanus, posterior sinus of 66
cavity of tympanus, promontory of 66
cavity of tympanus, sulcus of promontory of 66
cavity of tympanus, vestibular window of 66
cavity of uterus 79
cavity of uterus, entrance of oocyte into 66
cavity, axillary 8
cavity, buccal 31
cavity, filled with follicular fluid 84
cavity, nasal 31
cavity, region of axillary 8
cavity, thoracic 87
cavity, tympanic 66
cecum 39
cecum, coronal section of 39
cell 10, 11
cell arranged in concentric layers 80
cell of a taste bud 69
cell of Schwann 29
cell, centriole of the 10
cell, chromatin of eukaryotic 11
cell, eukaryotic 11
cell, human 10
cell, mastoid 64
cell, mastoid of tympanus 66
cell, microfilaments of the 10
cell, microscopic view of a blood 51
cell, microscopic view of an hepatic 36
cell, microscopic view of pancreatic anciar 34
cell, microtubules of the 10
cell, microvilli of the 10
cell, nuclear membrane of a eukaryotic 11
cell, nuclear membrane of the 10
cell, nuclear pore of a eukaryotic 11
cell, nuclear pore of the 10
cell, nucleolus of the 10
cell, nucleus of a eukaryotic 11
cell, nucleus of an olfactory 67
cell, nucleus of the 10
cell, olfactory 67
cell, pigmentary, of retina 59
cell, plasma membrane of a eukaryotic 11
cell, prokaryotic 11
cell, sensitive 70
cell, support cell or olfactory membrane 67
cell, support cell or olfactory mucosa 67
cell, taste 70
cell, tympanic 66
cell, types of 11
cells of epidermis, stratum spinosum 71
cells, layer of ganglion 59
cementum 41
cementum of a premolar bone 42
central of retina, artery 59, 60
centriole of the cell 10
cephalic vein 47
cerebral aqueduct 63
cervical ganglion, superior 62
cervical vertebrae 21
cervix of uterus 79, 88
chain formed by sugar/ phosphate 11
chain of glucose molecules 11
chamber of eye, anterior 60
chamber of eye, posterior 60
cheek 6, 7
chiasm, optic 54, 62, 63
chin 6, 7
choledochus, ductus 34, 35, 37
chorda tympani 64
chordae tendinae 68
chorion 86, 89
chorion, villi of 89
chorionic plate 86
choroid 59, 60, 62
choroid plexus of IV ventricle 54
choroid plexus of lateral ventricle of brain 63
chromatin of a eukaryotic cell 11
ciliary artery, long posterior 59
ciliary artery, posterior 60
ciliary artery, short posterior 59
ciliary muscle 60, 62
ciliary muscle of eye 59
ciliary muscle, long 61
ciliary nerve 62
cingulate sulcus 53
circular muscle 29
circumvallate papilla 68, 69, 70
clavicle 6, 7, 13, 14, 15, 16, 19, 26
clitoris 79, 88
clitoris, body of 79
clitoris, corpus of 74
clitoris, frenullum of 79
clitoris, glans of 74, 79
clitoris, prepuce of 79
cluneal nerve, inferior 58
cluneal nerve, superior 58
coccyx bone 13, 15, 16, 21, 22, 58, 88
coccyx nerve 58
cochlea 65, 66
cochlea, base of 65
cochlea, cupula of 65
cochlea, duct of 65
cochlea, fossa of window of 64, 66
cochlea, osseous 65
cochlea, perilymphatic space of 65
cochlea, scala tympani of osseous 65
cochlea, scala vestibuli of osseous 65
cochlea, window of 65
cochlear nerve 64, 65, 66
colic artery, middle 35
colic flexure, left 39
colic flexure, right 39
colic vein, middle 35
colic vein, right 35
colliculus seminal of prostatic portion of male urethra 77
colon 39
colon, anterior view of 39
colon, ascending 31, 39
colon, circular layer of tunica muscularis 39
colon, coronal section of ascending 39
colon, descending 31, 39
colon, imprint on liver of 39
colon, intestinal gland of mucosa 39
colon, lamina propria of mucosa 39
colon, longitudinal layer of tunica muscularis 39

colon, lymphatic nodule of
mucosa of 39
colon, mucosal epithelium 39
colon, muscularis mucosa of 39
colon, paracolic sulcus of 39
colon, projection on anterior
abdominal wall 39
colon, schematic section of
wall of 39
colon, semilunar fold of 39
colon, serosa of 39
colon, sigmoid 31, 78, 79, 88
colon, submucosa of 39
colon, transverse 31, 39
column of Bertin (renal) 75
commisure of female genitalia,
anterior labial 79
commisure of female genitalia,
posterior labial 79
common bile duct 34, 37
components of a four-week-old
embryo 87
components of the digestive
system 31
concha 64
concha of external ear, cavity of 64
concha of external ear, superior
fossa of 64
concha, inferior nasal 17, 67
concha, middle nasal 17, 67
concha, superior nasal 67
condyle of femur, lateral 15, 16
condyle of femur, medial 15
condyle of tibia, lateral 20
condyle of tibia, medial 20
condyle, occipital 17
cone of light of tympanic
membrane 64
cone of retina 59
cones and rods of retina, layer of 59
conjunctiva 59
conjunctiva of sclera 61
conjunctiva with Meibom glands,
inferior eyelid 61
conjunctiva with Meibom glands,
superior eyelid 61
conjunctiva, ocular 60
contact structure of epidermis 71
contraction in labor 89
conus arteriosus 49
coracobrachialis muscle 24
cord of fetus, spinal 87
cord, umbilical 86, 88, 89
cornea 59, 60, 61, 63
cornea, endothelium of 59
cornea, posterior surface of 62
cornea, substancia propria of 59
corona of glans of penis 77, 78
corona of penis 78
coronary artery 47
coronary artery, circumflex branch
of left 49
coronary artery, right 48
coronary vein 49
corpus albicans 81
corpus callosum, body of 63
corpus cavernosum of penis 74, 78
corpus cavernosum of penis,
trabeculae of 78
corpus cavernosus of male
urethra 77
corpus cavernosus of urethra 77
corpus cavernosus of urethra,
trabeculae of 77
corpus cavernosus of urethra,
tunica albuginea of 77
corpus luteum 81, 82, 83, 84, 85
corpus luteum during
menstruation 79
corpus of clitoris 74
corpuscle of Krause 71

corpuscle of Vater-Pacini 71
corpuscle, touch 71
corrugator supercilii muscle
(of eyebrow) 27
cortex, cerebral 70
cortex, optic field on occipital
lobe 62
cortex, renal 75
Corti, spiral organ of 65
corticose vein 59, 60
crest of gluteus maximus muscle 15
crest of greater tubercle of
humerus 19
crest of lesser tubercle of
humerus 19
crest of nail matrix 73
crest of prostatic portion of male
urethra, 77
crest of pubis 20
crest, external occipital 17
crest, frontal 17
crest, iliac 16, 25
crest, lateral supracondylar crest
of humerus 19
crest, medial supracondylar crest
of humerus 19
crest, median of sacrum bone 22
crest, obturator of os coxae bone 20
crest, posterior intertrochanter 15
crest, urethral 77, 78
cricoid cartilage, arch of 45
cristae, mitochondrial 10
crown of incisor tooth 41
crown of premolar tooth 42
crown of tooth 41
crural muscle 24
crus of clitoris 79
crus of greater alar cartilage of
nose, lateral 67
crus of incus, inferior 66
crus of incus, superior 66
crus of incus, vertical 66
crus of osseous labyrinth,
common 65
crus of penis 77, 78, 80
crus of stapes, anterior 66
crus of stapes, long 64
crus of stapes, posterior 66
crypt, tonsilar 68
cubital artery 47
cubital nerve 55, 58
cubital region 8, 9
cuboid muscle 16, 18
cuneiform bone, intermediate 16, 18
cuneiform bone, lateral 16, 18
cuneiform bone, medial 16, 18
cupola of stomach 32
cupula of cochlea 65
curvature of cristaline lens 63
curvature of stomach, greater 32, 33
curvature of stomach, lesser 32, 33
curve, thermal 84
cusp of left atrioventricular valve,
right commissural 48
cusp of mitral valve, left
commissural 48
cusp of mitral valve, right
commissural 48
cusp of right atrioventricular valve
(tricuspid), anterior 48
cusp of right atrioventricular valve
(tricuspid), posterior 48
cusp of right atrioventricular valve
(tricuspid), septal 48
cusp of right atrioventricular valve
of posterior 48
cusp of right atrioventricular valve
of septal 48
cusp of tricuspid valve of
posterior 48
cusp of tricuspid valve of septal 48

cusp, anterior pulmonary
semilunar 48
cusp, aortic left semilunar 48
cusp, aortic right semilunar 48
cusp, aortic valve of left
semilunar 48
cusp, aortic valve of posterior
semilunar 48
cusp, aortic valve of right
semilunar 48
cusp, left atrioventricular
commissural 48
cusp, left atrioventricular
posterior 48
cusp, left atrioventricular valve of
anterior 48
cusp, left atrioventricular valve of
left commissural 48
cusp, left atrioventricular valve of
posterior 48
cusp, left pulmonary semilunar 48
cusp, mitral left commissural 48
cusp, mitral valve of anterior 48
cusp, posterior aortic semilunar 48
cusp, right aortic semilunar 48
cusp, right atrioventricular
posterior 48
cusp, right atrioventricular septal 48
cusp, right atrioventricular valve of
anterior 48
cusp, right commissural valve of
mitral valve of anterior 48
cusp, right pulmonary semilunar 48
cusp, tricuspid posterior 48
cusp, tricuspid septal 48
cusp, tricuspid valve of anterior 48
cutaneous nerve, antebrachial 58
cutaneous nerve, anterior
femoral 55
cutaneous nerve, lateral
femoral 55, 58
cutaneous nerve, medial
antebrachial 58
cutaneous nerve, medial
brachial 55, 58
cutaneous nerve, posterior
antebrachial 58
cutaneous nerve, posterior
femoral 58
cuticle 73
cuticle of hair 72
cycle, follicular phase of
menstrual 84
cycle, menstrual 84
cycle, phase of proliferation of
menstrual 84
cymba of concha of external ear 64
cystic artery 37
cytoplasm 10, 11
cytoplasm of an eukaryotic cell 11
cytosine 11

D
decidua basale 86
deciduous tooth 41
decussation of pyramid of brain 54
defect, visual 63
deltoid muscle 24, 25
deltoid region 8, 9
dendrite of a bipolar neuron 56
dendrite of a motor neuron 56
dendrite of a multipolar neuron 56
dendrite of a unipolar neuron 56
dens (odontoid process) of axis 22
dental artery, inferior 41
dental nerve, inferior 41, 42
dental nerve, posterior 42
dental vein 41
dentina 41, 42
depressor anguli muscle
(of mouth) 27

depressor labii inferioris muscle
(of lip) 27
depressor septi nasi muscle
(of nose) 27
dermis 71
dermis, fat subungueal 73
Descemet, membrane of 59
descent of egg through the
Fallopian (uterine) tube 84
destruction of erythrocytes 31
detritus 31
development of auditory process 66
development of fetus 86
diagram of a nephron 76
diagram of bronchopulmonary
segments 44
diagram of visual pathways 62
diaphragm, (abdominal) 35, 36,
46, 88
diaphysis of clavicle 19
diaphysis of femur 20
diaphysis of fibula 20
diaphysis of humerus 19
diaphysis of metatarsal bone 18
diaphysis of phalanx of finger 18
diaphysis of phalanx of toe 18
diaphysis of tibia 20
diastole, atrial 50
diastole, ventricular 50
digestion of nutrients 31
digestion, phases of 31
digestive system 31–42
digestive system, organs of the 31
digital artery 47
digital vein 47
disc, intervertebral 14, 16
disc, optic 60
division of egg into blastomeres 83
division of egg into eight
blastomeres 82
division of egg into four
blastomeres 83
division of egg into two
blastomeres 83
DNA 11
dorsal artery of foot 47
dorsal nerve 58
dorsal scapular nerve 55
dorsal tongue muscles 68, 69
dorsum of penis 80
dorsum of the foot 8, 9
dorsum of the hand 8
duct of bulbourethral gland 74,
77, 78
duct of Muller 74
duct of Santorini 34
duct of Wirsung (pancreatic) 34
duct, aberrant 80
duct, anal 31
duct, bile 37, 64
duct, cochlear 65, 66
duct, common bile 37
duct, common hepatic 35, 37
duct, cystic 31, 35
duct, ejaculatory 74
duct, facial duct of tympanic
cavity 66
duct, hepatic 36
duct, inferior aberrant 74
duct, infundibulum of
paramesonephric (Mullerian) 74
duct, lactiferous 90
duct, left hepatic 37
duct, longitudinal of epoophoron 74
duct, mesonephric (Wolffian) 74
duct, nasolacrimal 61
duct, opening of pancreatic 34
duct, pancreatic (of Wirsung) 34, 35
duct, paramesonephric
(Mullerian) 74
duct, right hepatic 37

duct, section of sweat 72
duct, sweat 72
duct, Wolffian 74
ducts, tranverse of epoophoron 74
ductule, opening of ejaculatory 77
ductule, opening of prostatic 77
ductus deferens 74
ductus deferens of seminal
vesicle 78
ductus deferens of testis 78, 80
duodenal bulb 32
duodenal imprint on liver 37
duodenojejunal flexure 34
duodenum 31, 32, 33, 34, 35
duodenum, anterior view 32, 34
duodenum, ascending part of 34
duodenum, descending
part of 32, 34, 35
duodenum, general view 32
duodenum, inferior flexure of 35
duodenum, longitudinal
part of 35, 37
duodenum, lymphatic nodule of 37
duodenum, major papilla
(of Vater) of 37
duodenum, minor papilla of 37
duodenum, plica circularis of 31,
32, 34, 37
duodenum, posterior view 35
duodenum, projection on anterior
abdominal wall 34
duodenum, superior flexure of 35
duodenum, superior longitudinal
part of 34
duodenum, superior part of 32, 35
duodenum, suspensory muscle
of 34, 35
dura mater 59, 60

E
ear, antihelix of 64, 66
ectoderm 85
egg 81, 85
egg division in blastomeres 83
egg transformed into blastula 83
egg, arrival at uterus of 84
egg, descent through the Fallopian
(uterine) tube 84
egg, formation of 83
egg, stay in uterine cavity of free 84
elastic laminar, external 52
elastic laminar, internal 52
elbow 6, 7
elbow, anterior region of 8
elbow, posterior region of 9
embryo, components of a four-
week 87
eminence, iliopubic 20
eminence, intercondylar 14
eminence, parietal 17
enamel of tooth 41, 42
end plate, motor 73
ending, free nerve 56
ending, sensory nerve 71
ending, types of nerve 56
endoderm 85
endometrium 79, 85
endomysium 29
endoplasmic reticulum, rough 10
endoplasmic reticulum, smooth 10
endothelium of a vein 52
endothelium of an artery 52
endothelium of cornea 59
entrance of oocyte into the uterine
cavity 82
entrance of oxygen (O_2) 46
eosinophil 51
epicardium 48
epicondyle of femur, lateral 14, 15,
16, 20
epicondyle of femur, medial 20

epicondyle of humerus, lateral 19
epicondyle of humerus, medial 19
epicondyle, humeral 14, 15, 25
epicranius muscle 27
epicranius muscle, frontal
belly of 27
epicranius muscle, occipital
belly of 27
epidermis 71, 72, 73
epidermis, contact structure of 71
epidermis, stratum corneum of 71
epidermis, stratum germinatium
of 71
epidermis, stratum granulosum
of 71
epidermis, stratum lucida of 71
epidermis, stratum spinosum of 71
epididymis 78, 80
epididymis, appendix of 80
epididymis, body of 80
epididymis, head of 74, 78, 80
epididymis, tail of 74, 78, 80
epigastric region 8
epigastric vessels 35
epiglottis 31, 68
epiphysis of humerus 19
epithelium of a tooth 41
epithelium of alveolus 46
epithelium of cornea 60
epithelium of cornea, anterior 59
epithelium of olfactory mucosa 67
epithelium of retina, pigmented 59
epithelium of tunica mucosa of
colon wall 39
epithelium of tunica mucosa of
ileal wall 38
epithelium of tunica mucosa of
stomach wall 33
epithelium, germinal 81
epithelium, olfactory 67
epitrochlea of humerus 14, 24
epoophoron 74
epoophoron, longitudinal ducts
of 74
epoophoron, transverse ducts of 74
equator of the eye 60
erector penis 78, 79
eruption age of teeth 41, 42
erythrocyte, destruction of 31
escision in the perineum 89
esophagus 31, 88
esophagus, abdominal section of
the 32
esophagus, angular notch of 31
esophagus, imprint on liver of
the 37
esophagus, muscular layer of
the 33
ethmoid bone 12, 16, 17
Eustachian tube 65
exhalation 46
exit of fetus to the exterior 89
expulsion of oocyte 82, 83, 85
expulsion of secondary oocyte 84
extensor carpi radialis brevis
muscle 26
extensor carpi radialis longus
muscle 26
extensor communis digitorum
muscle (of finger) 25
extensor communis digitorum
muscle (of toe) 24, 28
extensor communis digitorum
muscle (of toe), tendon
of the 28
extensor digiti minimi muscle
(of little finger) 25
extensor digitorum brevis muscle
(of finger) 26
extensor digitorum brevis muscle
(of toe) 28

Index

extensor digitorum longus muscle (of toes) 28
extensor digitorum longus muscle (of toes), tendon of 28
extensor digitorum longus muscle 26
extensor digitorum longus muscle, tendon of 26
extensor digitorum longus of little finger 26
extensor digitorum muscle 26
extensor digitorum muscle, tendon of 26
extensor hallucis brevis muscle (of toe) 26
extensor hallucis longus muscle (of toe) 24
extensor hallucis longus tendon (of toe) 28
extensor indicis muscle (of index finger) 25
extensor pollicis brevis muscle (of thumb) 25, 26
extensor pollicis brevis muscle (of thumb), tendon of 28
extensor pollicis longus muscle (of thumb) 25
extensor, inferior retinaculum of 26
extensor, superior retinaculum of 24
external abdomen oblique, aponeurosis of 26
external acoustic meatus 16, 17, 64, 66
external aggressions, response of the body to 73
extremity of clavicle, acromial 19
extremity of clavicle, sternal 19
extremity of spleen, anterior 40
extremity of spleen, posterior 40
eye 59–63
eyebrow 61
eyebrow of fetus 87
eyebrow, corrugator supercilii muscle of 27
eyelash 61
eyelashes of fetus 87

F

face 6, 7
facet for transverse ligament of atlas 22
facet of symphysis pubis 20
facet of the sixth thoracic vertebra, inferior costal 22
facet of the sixth thoracic vertebra, superior costal 22
facet, anterior articular of axis 22
facet, articular costal 21
facet, articular of axis for dens 23
facet, inferior articular of atlas 23
facet, inferior articular of axis 22
facet, inferior articular of fifth lumbar vertebra 23
facet, small inferior costal of thoracic vertebra 22
facet, small superior costal of the sixth thoracic vertebra 22
facet, superior articular of atlas 23
facet, superior articular of axis 22
facet, superior articular of fourth cervical vertebra 23
facet, superior articular of the seventh cervical vertebra 23
facet, transverse costal of fifth thoracic vertebra 22
facial nerve 54, 70
Fallopian tube 79, 82, 83, 85
Fallopian tube, infundibulum of the 74, 79, 82, 83
Fallopian tube, path of oocyte toward uterus through the 82
fascia, parotid 27

fascia, umbilical prevesical 78
fasciculus, lateral 55
fasciculus, middle 55
fasciculus, posterior 55
fat 90
fat of the skin 71
fat, intestinal absorption of 31
femoral anterior region 8
femoral artery 47
femoral nerve 55
femoral posterior region 9
femoral region 8
femoral vein 47
femur 13, 14, 15, 20
femur, bone 13, 14, 15, 16, 20
femur, head of 16
femur, lateral condyle of 16
femur, lateral epicondyle of 16
fertilization 84, 85
fertilization of the oocyte by one spermatozoon 83, 85
fertilization, mechanism of 84
fetal development in mother's womb 86
fetus 86, 88, 89
fetus during gestation 87
fetus, arm of the 87
fetus, birth of the 89
fetus, blood of the 87
fetus, bone marrow of the 87
fetus, brain of the 87
fetus, capillaries of the 87
fetus, digestive system of the 87
fetus, eighth month of the 87
fetus, eyebrows of the 87
fetus, eyelashes of the 87
fetus, fifth month of the 87
fetus, fingernail of the 87
fetus, first month of the 87
fetus, foot of the 87
fetus, fourth month of the 87
fetus, fourth week of the 87
fetus, hair of the 87
fetus, hand of the 87
fetus, head of the 89
fetus, human resemblance of 87
fetus, internal organs of the 87
fetus, kidney of the 87
fetus, leg of the 87
fetus, liver of the 87
fetus, lung of the 87
fetus, nervous system of the 87
fetus, ninth month of the 87
fetus, pancreas of the 87
fetus, red blood cell of the 87
fetus, second month of the 87
fetus, seventh month of the 87
fetus, sexual organs of the 87
fetus, sixth month of the 87
fetus, spinal cord of the 87
fetus, third month of the 87
fetus, third week of the 87
fetus, thorax of the 87
fiber of a main artery, collagen 52
fiber of a main artery, elastic 52
fiber of a main vein, collagen 52
fiber of a main vein, elastic 52
fiber of alveolus, elastic 46
fiber of an artery, muscular 52
fiber of an olfactory cell, nerve 67
fiber of brain, primary 70
fiber of brain, secondary 70
fiber of brain, tertiary 70
fiber of muscular layer of colic wall, longitudinal 38
fiber of muscular layer of iliac wall, longitudinal 38
fiber of muscular layer of stomach, longitudinal 33
fiber of muscular layer of wall of stomach, longitudinal 33

fiber of muscular layer of wall of stomach, muscular 38
fiber of muscular wall of colon, muscular 39
fiber of muscular wall of ileum, muscular 38
fiber of muscular wall of stomach, muscular 33
fiber of stomach, longitudinal 33
fiber, layer of optic nerve 59
fiber, muscular 29
fiber, nucleus of a muscle 29
fiber, sympathetic bulb of a muscle 29
fiber, terminal motor plate of a muscle 29
fiber, zonular 62
fibrocartilagenous ring of tympanic membrane 64
fibula bone 13, 14, 15, 16, 20, 28
fibula, head of 24, 26
fibular artery 47
fifth lumbar vertebra, inferior articular facet of the 23
fifth lumbar vertebra, inferior articular process of the 23
fifth lumbar vertebra, inferior pedicle notch of the 23
fifth lumbar vertebra, lateral view of the 23
fifth lumbar vertebra, mammilary tubercle of the 23
fifth lumbar vertebra, pedicle of the 23
fifth lumbar vertebra, spinous process of the 23
fifth lumbar vertebra, transverse articular process of the 23
fifth lumbar vertebra, vertebral body of the 23
fifth metatarsal bone, tubercle of the 18
fifth thoracic vertebra 22
fifth thoracic vertebra, annular process of the 22
fifth thoracic vertebra, costal facet of transverse process of the 23
fifth thoracic vertebra, intervertebral face of the body of the 22
fifth thoracic vertebra, pedicle of the 23
fifth thoracic vertebra, spinous process of the 22
fifth thoracic vertebra, superior articular facet of the 22
fifth thoracic vertebra, transverse process of the 22
filiform papilla 70
filling of atrium 50
filtration of blood 76
finger 6, 7, 18, 19, 20
finger of hand, annular part of fibrous sheath of 28
finger of hand, base of phalanx 18
finger of hand, body of phalanx 18
finger of hand, cruciate part of fibrous sheath of 28
finger of hand, distal phalanx of 18
finger of hand, extensor digitorum comunis muscle of 25
finger of hand, flexor digitorum profundus muscle of 28
finger of hand, flexor digitorum superficialis muscle of 24
finger of hand, middle phalanx of 18
finger of hand, proximal phalanx of 18
finger of hand, section of a 78
finger of hand, tuberosity of distal phalanx of 18

finger, abductor digiti minimi muscle of little 28
finger, bone of 13, 73
finger, epidermic groove of 73
finger, extensor digitorum brevis muscle of 26
finger, extensor digitorum comunis muscle of 25
finger, extensor digitorum longus muscle of 26
finger, extensor digitorum muscle of 26
finger, extensor tendon of 26
finger, fibrous sheath of 24
finger, flexor digiti minimi muscle of little 28
finger, flexor digitorum longus muscle of 25
finger, little 18
finger, middle 18
finger, opponents digiti minimi muscle of little 28
finger, ring 18
finger, section of a 73
first cervical vertebra 16
fissure of brain, longitudinal 54
fissure of left lung, interlobar 43
fissure of left lung, oblique 43
fissure of ligamentum teres of liver 37
fissure of right lung, horizontal 43
fissure of right lung, oblique 43
flagellum of a spermatozoon 80
flexor digiti minimi muscle 28
flexor digitorum profundus muscle 24
flexor digitorum superficialis muscle 24
flexor digitorum longus muscle 25
flexor digitorum profundus muscle, tendon of the 28
flexor pollicis longus muscle 24, 25, 26
flexor pollicis longus tendon, sheath of 28
flexor pollicis muscle 28
flexor retinaculum 24, 25
flocculus of cerebellum 54
fluid, follicular cavity filled with follicular 84
fold of colon, semilunar 39
fold of duodenum, circular 32, 34, 37
fold of duodenum, longitudinal 37
fold of gastric mucosa (rugae) 31
fold of iris 62
fold of male urethra, mucous 77
fold of mucosa of gallbladder 37
fold of mucosa of urinary bladder 77
fold of stomach, gastric 32
fold of tongue, lateral glossoepiglottic 68
fold of tongue, triangular 68
fold of tympanic membrane, posterior malleolar 64
fold of urinary bladder, mucous 77
fold, anterior malleolar 64
fold, cephalic 87
fold, cilliary 62
fold, duodenal 31
fold, gluteal 6, 7
fold, intergluteal 6, 7
fold, interureteric 77
fold, median glossoepiglottic 68
fold, sublingual 68
fold, vaginal 79
foliate papilla 70
follicle Graafian 79, 81, 84
follicle of colic mucosa, lymphatic 39
follicle of the upper section of small intestine, solitary 38

follicle, developed 84
follicle, hair 71, 72
follicle, lingual 68
follicle, mature open 81
follicle, opening of the 82, 83, 84, 85
follicle, phase of growth of the 84
follicle, primary 81, 82, 83, 85
follicle, secondary 81, 82, 83, 85
follicle, section of a hair 72
follicles of colon, lymph 39
follicles of duodenum, lymph 37
follicles of gastric mucosa, lymph 33
foot 6, 7, 18, 20
foot of fetus 87
foot, base of phalanx of toe 18
foot, bones of 18
foot, diaphysis of phalanx of toe 18
foot, distal phalanx of toes 18
foot, dorsal interosseous muscle of 24
foot, dorsal view of 18, 28
foot, dorsum of 8, 9
foot, extensor communis tendons of 28
foot, extensor digitorum brevis muscle of 28
foot, extensor digitorum comunis tendon of 28
foot, extensor digitorum longus muscle of 24, 28
foot, extensor digitorum longus muscle of 28
foot, head of phalanx of toe 18
foot, intermediate cuneiform bone of 18
foot, lateral cuneiform bone of 18
foot, lateral view of 18, 28
foot, medial cuneiform bone of 18
foot, middle phalange of toes of 18
foot, muscles of 28
foot, phalanges of the 16
foot, phalanges of toes 18
foot, plantar region of 9
foot, proximal phalanx of toes 18
foramen cecum 68
foramen of sacrum bone 22
foramen, greater sciatic 16
foramen, lesser sciatic 16
foramen, nutrient 19
forearm (antebrachium) 6, 7, 19
forearm completely flexed 30
forearm half way down 30
forearm half way up 30
forearm, lateral antebrachial cutaneous nerve of 58
forearm, medial antebrachial cutaneous nerve of 58
forearm, median nerve 55
forearm, movements of 30
forearm, posterior antebrachial cutaneous nerve of 58
forefinger 18
forehead 6, 7
formation of morula 83, 85
formation of the egg 83
formation of the zygote 83
forms and types of muscles and muscular insertions 29
fornix conjunctiva, inferior 61
fornix conjunctiva, superior 61
fornix, posterior vaginal 79
fossa of brain, interpeduncular 54
fossa of cochlear window 64, 66
fossa of external ear, triangular 64
fossa of female urethra, navicular 77
fossa of ilium 20
fossa of incus 66
fossa of male urethra, navicular 77, 78
fossa of vagina, vestibular 79
fossa, infraspinatus 15, 19, 26

fossa, intercondylar 15
fossa, lateral view of nasal 67
fossa, nasal 67
fossa, olecranon 15
fossa, popliteal 6, 7, 9
fossa, region of the infraclavicular 8
fossa, supraspinatus 15
fossa, temporal 17
fossa, tonsilar 68
fossa, trochanteric 20
fourth cervical vertebra, bifid spinous process of 23
fourth cervical vertebra, hairpin of forked spinous process of 23
fourth cervical vertebra, lamina of 23
fourth cervical vertebra, pedicle of 23
fourth cervical vertebra, superior articular facet of 23
fourth cervical vertebra, superior articular surface of vertebral body of 23
fourth cervical vertebra, superior notch of pedicle of 23
fourth cervical vertebra, superior view of 23
fourth cervical vertebra, transverse foramen of 23
fourth cervical vertebra, transverse process of 23
fourth cervical vertebra, unciform process of 23
fourth cervical vertebra, vertebral body of 23
fourth cervical vertebra, vertebral foramen of 23
fourth lumbar vertebra 16, 23
fourth lumbar vertebra, accesory process of 23
fourth lumbar vertebra, costal process of 23
fourth lumbar vertebra, intervertebral surface of vertebral body of 23
fourth lumbar vertebra, lamina of 23
fourth lumbar vertebra, mamillary process of 23
fourth lumbar vertebra, margin of crest of 23
fourth lumbar vertebra, pedicle of 23
fourth lumbar vertebra, process of Genebaur of 23
fourth lumbar vertebra, spinous process of 23
fourth lumbar vertebra, superior articular process of 23
fourth lumbar vertebra, vertebral body of 23
fourth lumbar vertebra, vertebral foramen of 23
fovea centralis 59, 60
frenulum of clitoris 79
frenulum of labium 79
frenulum of tongue 62
frontal bone 12, 14, 16, 17, 67
frontal nerve 61
frontal region 8, 9
frontalis muscle 24, 26
fundus of gallbladder 37
fundus of stomach 32, 33
fundus of uterus 79, 82, 83, 85
fusion of Mullerian tubes 74
fusion of one ovum and one spermatozoon 85

G

galea aponeurotica 24, 25, 27
gallbladder 31, 36, 37
gallbladder, anterior view of 37
gallbladder, body of 37

gallbladder, bottom of 37
gallbladder, mucosa of 37
gallbladder, mucosal fold of 37
gallbladder, neck of 37
gallbladder, serosa of 37
gallbladder, spiral fold of 37
ganglion, ciliary 62
ganglion, of spinal nerve
    (dorsal root ganglion) 57
ganglion, paravertebral
    (sympathetic) 57
ganglion, superior cervical 62
ganglion, sympathetic 78, 79
ganglion, trigeminal 42, 54
gastric artery 47
gastric artery, left 35
gastric vein, left 35
gastrocnemius muscle 24, 25, 26
gastrocnemius muscle, external
    portion of the 25
gastrocnemius muscle, internal
    portion of the 25
gastroduodenal artery 35
gastroduodenal vein 35
gastroepiploic vein 47
Gegenbaur of the fourth lumbar
    vertebra, process of 23
genioglossus muscle 68
geniohyoid muscle 68
genitalia, female 7
genitalia, female external 79, 80
genitalia, female labial nerves of 79
genitofemoral nerve 55, 58
genitourinary system, lateral view
    of female 74
genitourinary system, lateral view
    of male 74
gestation, fetus during 87
gingival epithelium (gum) of
    incisor tooth 41
gingival epithelium (gum) of
    premolar tooth 41
gland of Bartolino 74
gland of Bowman 67
gland of colic mucosa intestinal 39
gland of mucosa or ileum,
    intestinal 38
gland of stomach mucosa 31
gland of tunica mucosa of wall of
    ileum 38
gland of Zeis 61
gland, anterior lingual 68
gland, bronchial mucosal 46
gland, bulbourethral (Cowper's) 74,
    77, 78
gland, duct of bulbourethral 74,
    77, 78
gland, gastric mucosal 33
gland, hilus of suprarenal 75
gland, imprint on liver of
    suprarenal 37
gland, lacrimal 61
gland, mammary 90
gland, sebaceous 71, 72
gland, suprarenal 35, 75
gland, sweat 71, 72
gland, tracheal 45
gland, uterine 84
glands, areolar 90
glands, greater vestibular
    (Bartholin's) 74
glans of clitoris 74, 79
glans of penis 74, 77, 78, 80
globe, ocular 62, 63
glomerulus 76
glomerulus, afferent artery of 76
glomerulus, renal 76
glossopharyngeal nerve 54, 70
glucose molecules, chain of 11
glucose, chain of molecules of 11

gluteal region 9
gluteus maximus muscle 25, 26, 79
gluteus medius muscle 24, 25
glycogen 11, 31
Golgi apparatus 10
Graafian follicle 84
granule, cortical 81
greater wing of sphenoid bone 17
groin 6, 7
groove of finger, epidermic 73
groove, lateral ophthalmic 54
growth phase of follicle 84
guanine 11
gum 41, 42
gyrus of brain, angular 53
gyrus of brain, inferior parietal 53
gyrus of brain, lateral
    occipitotemporal 54
gyrus of brain, middle frontal 53
gyrus of brain, occipital 53
gyrus of brain, orbital 54
gyrus of brain, postcentral 53
gyrus of brain, precentral 53
gyrus of brain, superior frontal 53
gyrus of brain, supramarginal 53
gyrus, hippocampal 54
gyrus, orbital 63

H
hair 6, 7
hair cuticle 72
hair of pubis 80
hair, basement membrane of a 72
hair, conjunctal tissue of a 72
hair, external root sheath of a 72
hair, matrix of a 72
hair, medulla of a 72
hair, papilla of a 72
hair, section of a 71
hairpin of bifid spinous process of
    the fourth cervical vertebra 23
hamate bone 14, 18, 28
hand 6, 7, 18, 19
hand of fetus 87
hand, annular part of fibrous
    sheath of fingers of 28
hand, back of 9
hand, base of phalanx of finger 18
hand, body of phalanx of finger 18
hand, bones of 13
hand, cruciate part of fibrous
    sheath of fingers 28
hand, distal phalanx of finger of 18
hand, dorsal interosseous muscle
    of 24
hand, dorsal view of the 18, 28
hand, dorsum of the 8
hand, lumbrical muscle of 28
hand, middle phalanx of finger of 18
hand, muscles of 28
hand, palm of the 8, 9
hand, palmar view of the 18, 28
hand, phalanx of 16
hand, proximal phalanx of finger
    of 18
hand, tendon of flexor digitorum
    profundus muscle of 28
hand, tuberosity of phalanx of
    finger 18
haustra of colon 31, 39
head 6, 7, 87
head of dorsal horn of spinal cord 73
head of epididymis 74, 78, 80
head of femur 15, 16, 20
head of fetus 89
head of fibula 14, 20, 24, 26
head of humerus 15, 19
head of malleus 64, 66
head of metacarpal bone 18, 28
head of metatarsal bone 18

head of pancreas 34, 35
head of phalanx of finger 18
head of phalanx of toe 18
head of radius 19
head of spermatozoon 80
head of talus 18
head of triceps brachii muscle,
    long 26
head of triceps brachii muscle,
    medial 26
head of triceps brachii muscle,
    short 26
head of ulna 19
head, adipose membrane of the 27
head, anterior view of the 27
head, bones of the 12
head, lateral view of the 27
head, muscles of the 27
head, semispinous muscle of the 27
head, splenic muscle of the 27
heart 46, 48, 49, 50, 87, 88
heart, anterior view of the 49
heart, apex of the 49
heart, cranial view of the valves
    of the 49
heart, posterior view of the 49
heart, projection on anterior
    abdominal wall of 48
heart, ventrolateral section of the 48
heel 6, 7
helix 64
helix canal 64
helix of ear 66
helix, crus of 64
hemisphere of cerebellum 54
Henle's loop 76
hepatic artery 31, 47
hepatic artery, common 35
hepatic artery, proper 35, 37
hepatic vein 47
hiatus of aorta 35
hilus of adrenal (suprarenal)
    gland 75
hilus of kidney 75
hilus of spleen 40
hip 6, 7
hook of triquetral bone 28
horn of spinal cord, dorsal 57
horn of spinal cord, ventral 57
horn of the fourth lumbar vertebra,
    margin of 23
humerus bone 13, 14, 15, 16, 19
hymen 79
hyoid bone 12, 16, 68
hypermetropia (far sightedness) 63
hypochondrial region 8, 9
hypodermis 71
hypogastric nerve 79
hypogastric nerve, inferior 79
hypogastric plexus 78
hypogastric plexus, inferior 78
hypogastric plexus, superior 78, 79
hypoglossal nerve 54
hypophysis of brain 54
hyppocampal gyrus 54

I
ileal artery 35
ileal vein 35
ileum 31, 39
ileum, circular fibers of the
    muscularis externus 38
ileum, epithelium of 38
ileum, intestinal gland of mucosa
    of 38
ileum, intestinal villi of 38
ileum, lamina propria of mucosa
    of 38
ileum, longitudinal fibers of the
    muscularis externus of 38

ileum, main lymphatic vessel of
    mucosa of 38
ileum, muscularis mucosa of 38
ileum, schematic section of wall
    of 38
ileum, serosa of 38
ileum, submucosa of 38
ileum, terminal part of 31
iliac artery, common 47
iliac artery, external 47
iliac artery, internal 47
iliac muscle 24
iliac spine, anterior inferior 19
iliac spine, anterior superior 19
iliac vein, common 47
iliac vein, external 47
iliac vein, internal 47
iliocostalis muscle 25
iliohypogastric nerve 55, 58
ilioinguinal nerve 58
iliotibial tract 25
ilium bone 13, 14, 20
implantation 85
implantation in uterus, phases of
    the genetic material (oocite +
    spermatozoon) until its 83
imprint of adrenal (suprarenal)
    gland on liver 37
imprint of colon on liver 37
imprint of costoclavicular
    ligament 19
imprint of duodenum on liver 37
imprint of esophagus on liver 37
imprint of kidney on liver 37
imprint of stomach on liver 37
imprint of tooth root on
    mandible 41, 42
imprint of tooth root on
    maxilla 41, 42
imprint of trigeminus 66
impulse, nerve 66, 70
incisor tooth 41, 42
incus 12
incus bone 65, 66
incus, body of 66
incus, fossa of 66
incus, long crus of 66
incus, short crus of 66
index finger 18
index finger, extensor indicis
    muscle of 25
inferior lingula of bronchial
    pulmonary segments 44, 45
inframammary region 8
infraorbital nerve 61
infrascapular region 9
infraspinatus muscle 25, 26
infundibulum of brain 54
infundibulum of Fallopian
    (uterine) tube 74, 79, 82, 83, 85
infundibulum of paramesonephric
    duct 74
infundibulum of uterine tube 74, 79,
    82, 83, 85
infundibulum, vascular 60
inguinal region 8
inhalation 46
innervation of female genital
    apparatus, spinal 79
innervation of male genital
    apparatus, spinal 79
innervation of teeth 42
innervation, motor 70
innervation, sensory 70
insertion of a muscle, broad 29
insertion of transverse ligaments
    of axis, tuberosity of atlas
    of the 23
insertion, muscular 29
interchange of gas in alveoli 46
intercostal muscle 26, 46

intercostohumeral nerve 58
intermediate vein 47
internal wall or endothelium of a
    capillary 52
interosseous muscle of foot,
    dorsal 24, 25
interosseous muscle of hand,
    dorsal 24
interosseous muscle, dorsal 28
interparietal bone 17
interscapular region 9
intertubercular sulcus of
    humerus 19
intestinal gland of mucosa of
    ileum 38
intestine, anterior view of large 39
intestine, circular fibers of
    muscularis externus of small 38
intestine, intestinal villi of mucosa
    on upper part of small 38
intestine, large 38, 39
intestine, lateral view of small 38
intestine, longitudinal fibers of
    muscularis externus of small 38
intestine, mesentery of small 38
intestine, muscular layer of small 38
intestine, muscularis mucosa of
    upper part of small 38
intestine, projection on anterior
    abdominal wall of large 39
intestine, projection on anterior
    abdominal wall of small 38
intestine, serosa of visceral
    peritoneum of small 38
intestine, small 31, 38, 88
intestine, solitary follicle of visceral
    peritoneum of small 38
intestine, submucosa of upper part
    of small 38
intestine, subserosa of visceral
    peritoneum of small 38
intestine, transverse section of a
    portion of small 38
intestine, visceral peritoneum of
    small 38
iris 59, 60, 61
iris, anterior view of 62
iris, fold of 62
iris, major ring of 62
iris, minor ring of 62
iris, posterior surface of 62
iris, posterior view of 62
iris, pupilar margin of 62
ischium bone 14
ischium bone, tuberosity of 16
isthmus of uterine tube 79
isthmus of uterus 79

J
jejunal artery 35
jejunal vein 35
jejunum 31, 34, 35
joint, incusmalleolar 66
joint, incustapedial 66
joint, interphalangeal 28
jugular vein, internal 47
juice, pancreatic 31
junction of Mullerian ducts 74
junction of spinous process of
    axis 22

K
kidney of fetus 87
kidney, 35, 74, 75
kidney, adipose capsule of 75
kidney, anterior view of 75
kidney, cortex of 75
kidney, fibrous capsule of 75
kidney, functional unit of 76
kidney, hilus of 75
kidney, medial margin of 75
kidney, schematic anterior
    view of 75

kidney, schematic section of left 75
kidney, superior margin of 75
knee 6, 7, 20
knee, anterior region of 8
knee, posterior region of 8
knee, region of the 8
Krause, corpuscle of 71

L
labial nerve of female genitalia 79
labial region 8
labium majora 79, 88
labium minora 79, 88
labyrinth of ear, anterior osseous
    ampulla of osseous 65
labyrinth of ear, anterior
    semicircular canal of osseous 65
labyrinth of ear, lateral
    semicircular canal of osseous 65
labyrinth of ear, posterior osseous
    ampulla of osseous 65
labyrinth of ear, posterior
    semicircular canal of osseous 65
labyrinth of ear, saccule of
    osseous 65
labyrinth of ear, transverse section
    of osseous 65
labyrinth of ear, utricle of
    osseous 65
labyrinth of ear, vestibular nerve of
    osseous 65
labyrinth, anterior osseous
    ampulla of osseous 65
labyrinth, common crus of
    osseous 65
labyrinth, lateral osseous ampulla
    of osseous 65
labyrinth, osseous 65
labyrinth, transverse view of
    osseous 65
labyrinth, vestibule of osseous 65
lacrimal canaliculus, inferior 61
lacrimal canaliculus, superior 61
lacrimal gland 61
lacrimal gland, nerve of 61
lacrimal organs 61
lacrimal papilla 61
lacrimal, bone 12, 16, 17
lacunae (of Morgagni) of male
    urethra 77
lacunae urethra of penis 78
lamina cribosa 60
lamina cribosa of sclera 60
lamina of a main artery, external
    elastic 52
lamina of a main artery, internal
    elastic 52
lamina of a main vein, external
    elastic 52
lamina of a main vein, internal
    elastic 52
lamina of cricoid cartilage 45
lamina of the fourth cervical
    vertebra 23
lamina of the fourth lumbar
    vertebra 23
lamina of the seventh cervical
    vertebra 23
lamina of thyroid cartilage 45
lamina of vertebral arch of axis 22
lamina propria of mucosa of
    colon 79
lamina propria of mucosa of
    ileum 38
lamina propria of mucosa of
    stomach, solitary follicle of 33
laryngopharynx 31
larynx 43, 45
larynx, aclitus of 68
larynx, anterior view of 45
larynx, superior aperture 68

larynx, ventricle of 68
lateral basal bronchial segments 44, 45
lateral bronchial segments 44, 45
lateral head of gastrocnemius muscle 25
lateral part of liver 36
lateral rectus muscle 59, 61
lateral region of the neck 8
lateral region of the thorax 8
latissimus dorsi muscle 25
layer of an artery, tunica adventitia 52
layer of an artery, tunica intima 52
layer of an artery, tunica media 52
layer of body of urinary bladder, muscular 77
layer of cones and rods of retina 59
layer of esophagus, muscular 32
layer of illeum, circular muscle of the muscularis externus 38
layer of illeum, longitudinal muscle of the muscularis media 38
layer of mucosa of upper part of small intestine, muscular 38
layer of mucosa of wall of ileum, muscular 38
layer of mucosa of wall of stomach, muscular 33
layer of optic fiber nerves 59
layer of small intestine, circular muscle of the muscularis externus 38
layer of small intestine, longitudinal muscle of the muscularis externus 38
layer of small intestine, muscle 38
layer of spinous cells, epidermis 71
layer of spleen, external 40
layer of stomach wall, muscle 33
layer of stomach, circular muscle of the muscularis externus 33
layer of stomach, longitudinal muscle of the muscularis externus 33
layer of stomach, muscle 33
layer of urinary bladder, muscle 77
layer, capillary cellular 52
layer, circular muscle of stomach 33
layer, circular muscle of wall of ileum 38
layer, corneal layer of epidermis 71
layer, external of an artery 52
layer, external plexiform 59
layer, germinative layer of epidermis 71
layer, granular layer of epidermis 71
layer, internal muscular of retina 59
layer, internal plexiform 59
layer, longitudinal muscle of ileum 38
layer, longitudinal muscle of muscularis externus of small intestine 38
layer, longitudinal muscle of muscularis externus stomach 33
layer, muscular fibers of wall of stomach 33
layer, muscular of mucosal of colon 39
layer, nuclear external of retina 59
layer, submucosa of superior portion of small intestine 38
layer, submucosa of wall of colon 39
layer, submucosa of wall of ileum 38
layer, submucosa of wall of stomach 33
layer, visceral peritoneum of small intestine 38
layer, wall of colon 39

layer, wall of stomach 33
left lumbar region 8, 9
left posterior renal artery 75
leg 6, 7, 20
leg of fetus 87
leg, anterior region of the 8, 9
leg, posterior region of the 8, 9
lens 60
lens of the eye 59, 60, 63
lens of the eye, curvature of 63
lens, anterior pole of 62
lens, convergent 63
lens, divergent 63
levator alaeque nasi labii superioris muscle 24
levator alaeque nasi muscle 27
levator anguli oris muscle 27
levator ani muscle 79
levator labii superioris alaeque nasi muscle 24
levator labii superioris muscle 24, 27
levator menti muscle 24
levator palpebrae superioris muscle 61
levator scapulae muscle 26
ligament of atlas, facet for transverse 22
ligament of axis, tuberosity of atlas for insertion of 23
ligament of liver, coronary 36, 37
ligament of liver, falciform 36
ligament of liver, left triangular 36
ligament of liver, right triangular 36
ligament of ovary, suspensory 79
ligament of penis, suspensory 78
ligament of trachea, annular 43, 45
ligament of uterus, round 74
ligament of vena cava 37
ligament, anterior annular 28
ligament, anterior annular tarsal 28
ligament, falciform 31
ligament, gastrosplenic 40
ligament, medial palpebral 61
ligament, median umbilical 77
ligament, middle cricothyroid 45
ligament, ovarian 79
ligament, patellar 24, 26
ligament, proper ovarian 74, 79
ligament, pubicprostatic 78
ligament, splenorenal 40
ligament, transverse carpal 28
ligament, transverse metacarpal 28
ligamentum arteriosum 49
ligamentum teres of liver 36, 37
ligamentum teres of liver, fissure of 37
ligamentum teres of liver, notch of 37
ligamentum venosum of liver 37
limb, bones of lower 20
limb, bones of upper 19
limb, lower 20
limb, upper 19
line of clavicle, trapezoid 19
line of iliac bone, intermediate 20
line of os coxae bone, intermediate 20
line of sacrum bone, transverse 22
line of Schwalde 59
line, inferior mouth 17
line, inferior nuchal 15
line, inferior temporal 17
line, posterior gluteal 16
line, superior mouth 17
line, superior nuchal 15
line, superior temporal 17
linea, alba 78, 79
lingual nerve 42
lingula of left lung 43
lingula, pulmonary segments inferior of 44, 45

lingula, pulmonary segments superior of 44, 45
link formed by sugar/phosphate 11
lip of ilium bone, external 20
lip of ilium bone, internal 20
lip of os coxae bone, external 20
lip of os coxae bone, internal 20
lip, depressor labii inferioris muscle of inferior 27
lip, inferior 68
lip, levator labii superioris muscle of 24, 27
lip, orbicularis oris muscle of 24, 26
lip, triangular muscle of 24
lipid 31
lips of female genitalia, anterior labial commisure of 79
lips of female genitalia, posterior labial commisure of 79
little finger 18
little finger, abductor digiti minimi muscle of 24, 28
little finger, flexor digiti minimi muscle of 28
little finger, opponens digiti minimi muscle of 28
liver 31, 36, 37, 88
liver of fetus 87
liver, anterior part of 36
liver, appendix fibrosa of 37
liver, caudate lobe of 37
liver, caudate process of 37
liver, coronary ligament of 36, 37
liver, diaphragmatic surface of 37
liver, dorsal view of 37
liver, falciform ligament of 36
liver, fissure of ligamentum teres of 37
liver, imprint of adrenal (suprarenal) gland on the 37
liver, imprint of colon on the 37
liver, imprint of duodenum on the 37
liver, imprint of esophagus on the 37
liver, imprint of kidney on the 37
liver, inferior margin of 36, 37
liver, lateral part of 36
liver, left lobe of 31, 36, 37
liver, left triangular ligament of 36
liver, ligamentum teres of 36, 37
liver, ligamentum venosum of 37
liver, middle part of 36
liver, notch of ligamentum teres of 37
liver, omental tubercle of 37
liver, papillary process of 37
liver, posterior part of 36
liver, projection on anterior abdominal wall of 36
liver, right lobe of 31, 36, 37
liver, right triangular ligament of 36
liver, ventral view of 37
lobar bronchus 43, 44
lobar bronchus, right upper 45
lobe of brain, cortex optic field on occipital 62
lobe of brain, inferior parietal 53
lobe of brain, striated area of occipital 62
lobe of brain, superior parietal 53
lobe of ear 66
lobe of left lung, inferior 43, 44
lobe of left lung, lateral surface of 43
lobe of left lung, superior 43, 44
lobe of liver, caudate 37
lobe of liver, diaphragmatic surface of right 36
lobe of liver, diaphragmatic surface of the left 36

lobe of liver, left 31, 37
lobe of liver, right 31, 37
lobe of right lung, inferior 43, 44
lobe of right lung, lateral surface of 43
lobe of right lung, middle 43, 44
lobe of right lung, superior 43, 44
lobule 64
lobule of testis 78, 80
lobules of mammary gland 90
location of permanent teeth 41
location of teeth in mouth 42
loin (back) 6, 7
long head of biceps femoris muscle 26
long muscle 29
longitudinal muscle of tongue, inferior 68
longitudinal muscle of tongue, superior 68
longitudinal muscle, superior 68
loop of a nephron, distal convoluted tubule of 76
loop of a nephron, proximal convoluted tubule of 76
loop of Henle 76
lumbar nerve 58
lumbar region 9
lumbar vertebrae 21
lumbodorsal fascia, 25
lumbrical muscles of hand 28
lunate bone 14, 18
lung 43, 46
lung of fetus 87
lung, base of 43
lung, cardiac notch of right 43
lung, diaphragmatic surface of 43
lung, horizontal fissure of right 43
lung, inferior lobe of left 43
lung, inferior lobe of right 43
lung, inferior margin of 43
lung, interlobar fissure of left 43
lung, lateral surface of superior lobe of right 43
lung, left 43
lung, lingula of left 43
lung, middle lobe of right 43
lung, oblique fissure of left 43
lung, oblique fissure of right 43
lung, projection of bronchii on the wall of the 43
lung, right 43
lung, superior lobe of left 43
lunule 73
lymphocyte 51
lymphopoiesis 31
lysosome 10

M

main visual defects 63
malleolus internal region 8
malleolus of tibia, medial 20
malleolus, lateral (external) 14, 15, 20, 25, 26, 28
malleolus, medial (internal) 15, 20, 25, 28
malleus bone 12, 65, 66
malleus, anterior process of 66
malleus, head of 64, 66
malleus, lateral process of 66
malleus, manubrium of 66
malleus, neck of 66
Malpighi, pyramid of 75
mammary region 8
man, anatomy of 6
mandible bone 12, 14, 15, 16, 31, 68, 87
mandible, imprint of teeth roots on 41, 42
mandible, section with triple root teeth 41

mandibular nerve 58
manubrium of malleus 64, 66
manubrium of sternum 14, 16
margin of fibula, anterior 20
margin of fibula, superior 20
margin of fourth lumbar vertebra crest 23
margin of glenoid cavities of atlas 59
margin of iris, pupilar 62
margin of kidney, medial 75
margin of kidney, superior 75
margin of liver, inferior 36, 37
margin of lung, inferior 43
margin of nail, free 73
margin of pupil 62
margin of radius, anterior 19
margin of radius, interosseous 19
margin of scapula, lateral 19
margin of scapula, medial 19
margin of scapula, superior 19
margin of spleen, inferior 40
margin of spleen, superior 40
margin of tibia, anterior 16, 20
margin of tibia, interosseous 20
margin of tongue 68
margin of ulna, interosseous 19
mass of a triple root tooth, cortical 41
mass of atlas, lateral 23
masseter muscle 24, 26
mastication, muscles involved in 41
mastoid antrum of tympanic cavity 66
material, genetic 11
matrix of a hair 72
matrix, crest of nail 73
matter of spinal cord, white 73
maxilla 12, 14, 16, 17, 31, 42, 61, 67
maxilla, frontal process of 67
maxilla, imprint of teeth roots on 41, 42
maxillary nerve 54
maxillary nerve, superior 42, 58
meatus, inferior nasal 67
meatus, middle nasal 67
meatus, superior nasal 67
mechanism of breathing 46
mechanism of fertilization 84
mechanism of pregnancy 84
medial antebrachial cutaneous nerve 55
medial basal bronchial pulmonary segments 44, 45
medial bronchial pulmonary segments 44, 45
medial head of gastrocnemius muscle 25
median nerve 55, 58
median vein, inferior 47
medulla (or pyramid of Malpighi), renal 75
medulla oblongata 54, 70
meeting of oocyte with spermatozoa 84, 85
Meissner corpuscle 56, 71, 72
Meissner's corpuscle 56, 71
melanocyte 72
membrane (or membrane of Shrapell), pars flaccida of tympanic 64
membrane of a cell, nuclear 10
membrane of a eukaryotic cell, nuclear 11
membrane of a hair, basal 72
membrane of an artery, elastic 52
membrane of Bowman's capsule, basal 76
membrane of Descemet 59
membrane of ovary, vitelline 81
membrane of retina, external limiting 59

membrane of retina, internal limiting 59
membrane of Shrapell 64
membrane of tympanus 54, 66
membrane, cone of light of tympanic 64
membrane, cytoplasmic 10
membrane, fibrocartilagenous ring of tympanic 64
membrane, hyaloid 60
membrane, olfactory 67
membrane, pars tensa of tympanic 64
membrane, plasma 10
membrane, posterior malleolar fold of tympanic 64
membrane, support cell of olfactory 67
membrane, tympanic 64
menses 84
menses, fertile 84
menstrual cycle, corpus luteum during 79
mental nerve (of chin) 42
mental region (of chin) 8
mentalis muscle 24, 27
Merkel's disc 56
mesenteric artery, inferior 35, 47
mesenteric artery, superior 35, 47
mesenteric of small intestine 38
mesenteric vein, inferior 35, 47
mesenteric vein, superior 35, 47
mesoderm 85
metabolism 31
metabolized food 31
metacarpal bone 16, 18, 19
metacarpal bone, base of 18
metacarpal bone, head of 18, 28
metacarpal bone, transverse ligament of 28
metatarsal artery 47
metatarsal bone 16, 18, 20
metatarsal bone, base of 18
metatarsal bone, body of 18
metatarsal bone, head of 18
metatarsal bone, tubercle of 18
metatarsal bone, tubercle of fifth 18
microfilament of cell 10
microscopic view of a blood cell 51
microscopic view of acinar cells of pancreas 34
microscopic view of hepatic cell 36
microscopic view of interior wall of stomach 32
microtubule of cell 10
microvilli of cell 10
midbrain 63
midbrain (cerebral) aqueduct 63
midbrain top 63
middle finger 18
middle part of liver 36
mitochondrion 10
mitral valve 48
molar tooth 41, 42
monocyte 51
mons pubis 79
morula 85
morula with blastomeres in blastula 83
morula, formation of 83, 85
morula, nesting in uterine wall 83
mother, womb of 89
mouth 6, 7, 87
mouth, depressor anguli muscle of 27
mouth, levator anguli oris muscle of 27
mouth, location of teeth in 27
mouth, marginal portion of orbicularis oris muscle of 27
mouth, orbicularis oris muscle of 27

movements of forearm 30
mucosa 82
mucosa muscularis 33
mucosa of gallbladder, fold of 37
mucosa of stomach, gland of 31
mucosa of upper part of small intestine, muscularis 38
mucosa of urinary bladder, fold of 37
mucosa of wall of illeum, lamina propria of 38
mucosa of wall of illeum, muscularis 39
mucosa of wall of stomach, solitary lymphatic follicles of 33
mucosa, epithelium of olfactory 67
mucosa, lamina propria of 39
mucosa, lingual 68
mucosa, schematic section of olfactory 67
mucosa, support cells of olfactory 67
mucosa, surface of olfactory 67
Muller, duct of 74
Muller, joint of ducts of 74
muscle (of Santorini), risorius 24, 27
muscle canal, tensor tympanic 66
muscle of duodenum, suspensory 34, 35
muscle of eyes, ciliary 59
muscle of foot, dorsal interosseous 24
muscle of hair, erector pilorum 71
muscle of hand, dorsal interosseous 24, 25
muscle of lip, triangular 24
muscle of pupil, dilator 59
muscle of pupillary sphincter 59
muscle of tongue, inferior longitudinal 68
muscle of tongue, superior longitudinal 68
muscle of tongue, transverse 68
muscle of tongue, vertical 68
muscle of Treitz 34
muscle of uterus 89
muscle, abductor digiti minimi 24, 28
muscle, abductor pollicis 28
muscle, abductor pollicis longus 24, 25
muscle, adductor brevis 24
muscle, adductor longus 24
muscle, adductor magnus 24
muscle, alaeque nasi levator labii superioris 24
muscle, anterior papillary 48
muscle, aponeurosis of external oblique 26
muscle, auricularis anterior 24
muscle, auricularis posterior 24
muscle, auricularis superior 24, 25
muscle, biceps brachii 24, 26, 29
muscle, biceps femoris 25
muscle, bipenniform 29
muscle, brachialis 24, 25, 26
muscle, brachioradialis 24, 25
muscle, broad 29
muscle, broad insertion or aponeurosis of a 29
muscle, buccinator 24, 26
muscle, central tendon of bipenniform 29
muscle, ciliary 60, 62
muscle, circular 29
muscle, coracobrachialis 24
muscle, corrugator supercilii 24, 27
muscle, crural 24
muscle, deltoid 24, 25, 26
muscle, depressor anguli 27
muscle, depressor labii inferioris 27
muscle, depressor septi nasi 27
muscle, dorsal interosseous 26

muscle, extensor carpi radialis brevis 26
muscle, extensor carpi radialis longus 24, 25, 26
muscle, extensor carpi ulnaris 24
muscle, extensor communis digitorum 25
muscle, extensor digitorum 28
muscle, extensor digitorum brevis 26
muscle, extensor digitorum longus 24, 26, 28
muscle, extensor hallucis brevis 26
muscle, extensor hallucis longus 24, 25
muscle, extensor indicis 25
muscle, extensor pollicis brevis 25, 26
muscle, extensor pollicis longus 25
muscle, external abdominal oblique 24, 25
muscle, external anal sphincter 28
muscle, flexor carpi ulnaris 24
muscle, flexor digiti minimi 28
muscle, flexor digitorum profundus 24
muscle, flexor digitorum superficialis 24
muscle, flexor pollicis 26, 28
muscle, flexor pollicis longus 24
muscle, forms and types of 29
muscle, frontal belly 27
muscle, frontalis 24, 26, 27
muscle, gastrocnemius 26
muscle, genioglossus 68
muscle, geniohyoid 68
muscle, gluteus maximus 25, 26
muscle, gluteus medius 24, 25
muscle, gracilis 24
muscle, iliacus 24
muscle, iliocostalis 25
muscle, inferior belly of omohyoid 26
muscle, inferior oblique 61
muscle, inferior rectus 25, 59, 61
muscle, inferior retinaculum of peroneal 28
muscle, inferior tarsal 61
muscle, infraspinatus 25, 26
muscle, intercostal 25, 26
muscle, internal abdominal oblique 25
muscle, lateral head of gastrocnemius 25
muscle, lateral rectus 60, 61
muscle, latissimus dorsi 25
muscle, levator anguli oris 27
muscle, levator ani 79
muscle, levator labii superioris 24, 27
muscle, levator labii superioris alaeque nasi 24
muscle, levator menti 24
muscle, levator palpebrae superioris 61
muscle, levator scapulae 26
muscle, long 29
muscle, long ciliary 61
muscle, margin of orbicularis oris 27
muscle, masseter 24, 26
muscle, medial head of gastrocnemius 25
muscle, medial rectus 60
muscle, mentalis 27
muscle, middle tendon of bipenniform 29
muscle, mylohyoid 68
muscle, nasalis 27
muscle, occipital belly of epicranius 27

muscle, occipitalis 25, 26
muscle, omohyoid 24, 26
muscle, opponens digiti minimi 28
muscle, opponens pollicis 28
muscle, opponens pollicis longus 26
muscle, orbicularis oculi 26, 27, 61
muscle, orbicularis oris 24, 26, 27
muscle, palatoglossus 68
muscle, palatopharyngeus 68
muscle, palmaris brevis 24
muscle, palmaris longus 24
muscle, papebral part of orbicularis oculi 27
muscle, pectineus 24
muscle, pectoralis major 24, 26, 90
muscle, peroneus brevis 24, 25, 26
muscle, peroneus longus 24, 25, 26
muscle, plantaris 25, 26
muscle, polygastric 29
muscle, posterior papillary 48
muscle, procerus 27
muscle, process of a 29
muscle, pronator quadratus 28
muscle, pronator teres 24, 26
muscle, psoas major 24, 35
muscle, pyloric sphincter 31, 34
muscle, pyramidalis nasi 24, 27
muscle, quadratus lumborum 35
muscle, raphe of a 25
muscle, rectus abdominal 88
muscle, rectus femoris 24, 26
muscle, retinaculum of peroneal 26
muscle, rhomboideus major 25
muscle, risorius 26
muscle, sartorius 24, 25, 26
muscle, semimembranous 25, 26
muscle, semispinalis capitis 27
muscle, semitendinous 25
muscle, septi nasi depressor 27
muscle, serratus anterior 24, 26, 90
muscle, sheath of rectus abdominus 26
muscle, soleus 24, 25, 26
muscle, splenius 26
muscle, splenius capitis 25, 27
muscle, sternocleidomastoid 24, 25, 26, 27
muscle, sternohyoid 24
muscle, superior longitudinal 68
muscle, superior oblique 25
muscle, superior rectus 59, 61
muscle, superior retinaculum of peroneal 28
muscle, supinator 24
muscle, temporalis 26
muscle, tendon of extensor digitorum communis 28
muscle, tendon of extensor digitorum longus 26, 28
muscle, tendon of extensor pollicis brevis 28
muscle, tendon of flexor digitorum profundus 28
muscle, tendon of tensor tympany 64
muscle, tendon of tibialis anterior 26
muscle, tensor fasciae latae 24
muscle, tensor tympani 64
muscle, teres minor 25, 26
muscle, tibialis anterior 24, 26, 28
muscle, trachealis 15
muscle, trapezium 24, 25, 26, 27
muscle, triceps brachii 24, 25
muscle, uterine 89
muscle, vastus lateralis 24, 25, 26
muscle, vastus medialis 24, 26
muscle, zygomaticus major 24, 27
muscle, zygomaticus minus 24, 27
muscles involved in mastication 41
muscles of foot 28

muscles of hand 28
muscles of hand, lumbrical 28
muscles of head 27
muscles of human body 24, 25, 26
musculocutaneous nerve 55
myelin sheath 56
myelin sheath of a motor neuron 56
mylohyoid muscle 68
myocardium 48
myocardium of left ventricle 48
myocardium of right ventricle 48
myocardium, cranial view of 48
myofiber 29
myometrium 79, 85, 86
myopia (near sightedness) 63

N
nail 73
nail matrix, crest of 73
nail of fetus 87
nail, free margin of 73
nail, root of 73
nail, roundlet of 73
nasal bone 12, 14, 16, 17, 67
nasal region 8
nasalis muscle 27
nasociliary nerve 61
navicular bone 16, 18
neck 6, 7
neck of femur 15, 20
neck of fibula 20
neck of gallbladder 37
neck of humerus, anatomical 15, 19
neck of humerus, surgical 15, 19
neck of incisor tooth 41
neck of malleus 66
neck of posterior horn of spinal cord 73
neck of premolar tooth 42
neck of radius 19
neck of scapula 19
neck of uterus 79, 82, 83, 85, 89
neck of uterus, vaginal portion of 79
neck, anterior region of 8
neck, lateral region of 8
nephron 76
nephron, afferent artery of a 76
nephron, collecting tube of a 76
nephron, diagram of a 76
nephron, distal convoluted tubule of a 76
nephron, proximal convoluted tubule of a 76
nerve 41
nerve of a circumvallate papilla 69
nerve, abducens 54
nerve, accesory 54
nerve, antebrachial cutaneous 55, 58
nerve, axillary 55, 58
nerve, bronchial 46
nerve, cervical 58
nerve, ciliary 62
nerve, coccyx 57
nerve, cochlear 64, 65, 66
nerve, cubital 55, 58
nerve, dorsal 58
nerve, dorsal scapular 55
nerve, femoral cutaneous 56
nerve, great sciatic 55
nerve, greater auricular 58
nerve, greater occipital 58
nerve, inferior gluteal 58
nerve, inferior mental 41, 42
nerve, lateral femoral cutaneous 55, 58
nerve, lesser occipital 58
nerve, lesser sciatic 55
nerve, medial antebrachial cutaneous 55, 58
nerve, medial brachial cutaneous 55, 58

nerve, median 55
nerve, ophthalmic 42
nerve, posterior antebrachial cutaneous 58
nerve, posterior femoral cutaneous 58
nerve, posterior mental 41, 42
nerve, sciatic 55
nerve, superficial peroneal 55, 58
nerve, superior gluteal 58
nerve, ventral roots of spinal 54
nerves of an artery 52
nesting of morula in uterine wall 83
nesting of oocyte in uterine wall 82, 83
net of Hallen 80
neuron, axon 56
neuron, dendrite 56
neuron, motor 29
neuron, Nissal bodies 56
neuron, soma 56
nipple 7, 88, 90
Nissl bodies 56
Nissl bodies, bipolar neuron 56
Nissl bodies, multipolar neuron 56
Nissl bodies, unipolar neuron 56
node of Ranvier 56
nose 6, 7
nose, greater alar cartilage of 67
nose, lateral crus of greater alar cartilage of 67
nose, lesser alar cartilage of 67
nose, vestibule of 67
notch of ear, anterior 64
notch of ligamentum teres of liver 37
notch of lung, cardiac 43
notch of pedicle of fifth lumbar vertebra, inferior 23
notch of pedicle of fourth cervical vertebra, superior 23
notch of scapula, superior 19
notch of sixth thoracic vertebra, inferior vertebral 22
notch of stomach, angular 31, 32
notch of stomach, cardiac 32
notch of ulna, radial 19
notch of ulna, trochlear 19
notch, buccal 31
notch, greater sciatic 20
notch, intertragic 64
notch, pancreatic 34
notch, sphenoidal 14, 17
notch, sphenomaxillary 14
notch, supraorbital 16
notch, ungueal 73
nucleic acid 11
nucleus of trigeminal nerve, motor 54

O
oculomotor nerve, common 61
oculomotor nerve, external 54, 61
opening of follicle 82, 83, 84, 85
opening of Meibomian gland 61
opening of the Fallopian tube, external (abdominal) 79
opening of the vagina 79
ophthalmic nerve 42
opponens pollicis longus muscle 26
opponents digiti minimi muscle 28
optic disc 59
optic nerve 61, 62, 63
organic food, digestion of 31
organs of female genitalia 81
organs of male genitalia 80
oropharynx 31
os coxae bone 13, 20
ossicles of the ear 66
ossicles of the middle ear 66

P
Pacinian corpuscle 56, 71
pars flaccida or membrane of Shrapell (of tympanus) 64
pars tensa of tympanic membrane 64
part of a circumvallate papilla, schematic section of 69
part of aorta, ascending 48, 49
part of aorta, descending 49
part of aorta, thoracic 35
part of duodenum, ascending 34
part of duodenum, descending 32, 34, 35
part of fibrous sheath of fingers, annular 28
part of fibrous sheath of fingers, cruciate 28
part of interventricular septum, membranous 48
part of interventricular septum, muscular 48
part of male urethra, prostatic 77
part of male urethra, prostatic sinus of prostatic 77
part of male urethra, seminal colliculus of prostatic 77
part of male urethra, urethral crest of prostatic 77
part of orbicularis oculi muscle, orbital 24
part of orbicularis oculi muscle, papebral 27
part of retina, ciliary 62
part of spermatozoon, middle 80
part of stomach, pyloric 32
part of the tongue, section of 69
part of uterine neck, vaginal 79
part of uterus, cervix 79
patella 13, 14, 16, 20, 24, 26
patellar ligament 26
path, diagram of visual 62
path, inferior view of brain showing the visual 63
path, motor 73
path, olfactory 67
path, sensory 73
pathway of oocyte to the uterus through the Fallopian tube 82
pathway of unfertilized oocyte 82
pelvis 6, 7
pelvis, maternal 87, 89
pelvis, renal 74, 75
penis 6, 78
penis, body of 80
penis, bulb of 77, 78
penis, bulb of corpus spongiosum of 78
penis, corona of 78
penis, corona of glans 77, 78
penis, corpus cavernosum of 74, 78
penis, corpus spongiosum of 74, 77, 78
penis, crus of 77
penis, deep artery of 77, 78
penis, deep dorsal vein of 78
penis, dorsal artery of 7
penis, dorsum of 80
penis, glans 74, 77, 78, 80
penis, lacunae of urethra 78
penis, prepuce of 77, 78, 85
penis, superficial dorsal vein of 78
penis, suspensory ligament of 78
penis, trabeculae of corpus cavernosum of 78
penis, tunica albiginea of corpus cavernosum of 78
perception of taste on tongue, areas of 69
pericardial sinus, transverse 49
pericardium 49, 88

perimetrium 79, 85
perineal raphe 79
perineum, incision of in 89
perineum, tendon of anterior 24
period, fertile 84
periodontium 41
periorbit 61
peritoneum 31, 33
peritoneum of small intestine, visceral 38
peritoneum, serosa of 38
periungueal area, 73
permanent teeth 42
permanent teeth, location of 41
peroneal nerve, deep 58
peroneal nerve, superficial 55, 58
peroneal retinaculum, inferior 25
peroneal retinaculum, superior 25
peroneus longus muscle 24, 25
peroneus longus tendon 25
peroneus tertius muscle 24, 26
peroneus tertius tendon 26, 28
peroxisome 10
pers tensa of incus 64
phalanges 14, 15, 16
phalanges of foot 16
phalanges of hand 16
phalanges of toes 18
phalanx of fingers, base of the 18
phalanx of fingers, body of the 18
phalanx of fingers, distal 18
phalanx of fingers, head of the 18
phalanx of fingers, middle 18
phalanx of fingers, proximal 18
phalanx of fingers, tuberosity of the 18
phalanx of toes, body of the 18
phalanx of toes, distal 18
phalanx of toes, head of the 18
phalanx of toes, middle 18
phalanx of toes, proximal 18
phase of corpus luteum secretion 84
phase of growth of follicle 84
phase of growth of genetic material (oocyte + spermatozoon) until its implantation in the uterus 83
phase of menstrual cycle 84
phase of oocyte until its expulsion during the period 82
phase of the digestive process 31
phase of the menstrual cycle, follicular 84
phases of the heartbeat 50
pia matter 59, 60
pigment, bile 31
pisiform bone 18, 28
placenta 86, 88, 89
placenta of the fifth week of pregnancy 89
plantar artery, lateral 47
plantar region of foot 9
plantaris muscle 25, 26
plantaris tendon 25
plate of a multipolar neuron, motor end 56
plate of a muscular fiber, motor end 29
plate, internal tibial 14
plate, ungueal 73
platelet 51
platysma muscle 27
plexiform layer, external 59
plexiform layer, internal 59
plexus of seminal vesicle, deferential 78
plexus of testis 78
plexus, brachial 55, 58
plexus, cervical 55, 58
plexus, deferential 78
plexus, hypogastric 71
plexus, inferior dental 42
plexus, inferior hypogastric 79
plexus, lumbar 58
plexus, mastoid branch of superficial cervical 55
plexus, orbital nervous 61
plexus, ovarian 79
plexus, prostatic 78
plexus, pudendal 58
plexus, sacral 58, 79
plexus, subcutaneous 71
plexus, subpapillary 71
plexus, superior hypogastric 78, 79
plexus, supraclavicular branch of superficial cervical 58
plexus, supracromial branch of superficial cervical 55
plexus, testicular 78
plexus, uterovaginal 79
pole of brain, frontal 53, 54
pole of brain, occipital 53, 54
pole of brain, temporal 54
pole of eye, anterior 59
pole of eye, posterior 59
pole of lens, anterior 62
polygastric muscle 29
pons of brain 54
popliteal artery 47
pore of an eukaryotic cell, nuclear 11
pore of sweat gland 72
pore of the cell, nuclear 10
pore, schematic section of a sweat gland 71
portal hepatic vein 47
portal vein 35, 37
portion of duodenum, inferior longitudinal 34
portion of duodenum, longitudinal 35
portion of duodenum, superior 32, 35
portion of male urethra, membranous 78
portion of orbicularis oris muscle, marginal 27
portion of pharynx, oral 31
portion of sacrum bone, lateral 22
portion of small intestine, intestinal villi of mucosa of the superior 38
portion of small intestine, muscular layer of mucosa of the superior 38
portion of small intestine, solitary follicle of the superior 38
portion of small intestine, submucosa of superior 38
portion of small intestine, transverse section of the superior 38
portion of trachea, membranous 45
posterior antebrachial region 8, 9
posterior basal, bronchopulmonary segment 44, 45
posterior region of elbow 9
posterior region of leg 8, 9
posterior region of liver 36
posterior region of neck 9
posterior, bronchopulmonary segment 44, 45
postsulcal area of tongue 68
pouch, rectouterine (cul-de-sac of Douglas) 79
pre-menses 84
pregnancy, mechanism of 84
pregnancy, placenta in the fifth week of 89
premolar tooth 41, 42
premolar tooth, cementum of a 42
premolar tooth, central cavity of a 42
premolar tooth, conjunctive tissue of a 42
premolar tooth, crown of a 42
premolar tooth, cusp of a 42
premolar tooth, gingival epithelium of a 42
premolar tooth, neck of a 42
premolar tooth, pulp of a 42
premolar tooth, root canal of a 42
premolar tooth, root foramen of a 42
premolar tooth, root of a 42
premolar tooth, schematic longitudinal section of a 42
presbyopia 63
pressure and vibration, sensory receptor for 71
presternal region 8
presulcal area of tongue 68
procerus muscle 27
process of atlas, foramen of transverse 23
process of atlas, transverse 23
process of axis, foramen of transverse 23
process of axis, inferior articular 22
process of axis, junction of spinous 22
process of axis, spinous 22
process of blood oxygenation developed in pulmonary alveolus 46
process of delivery 89
process of digestion 31
process of digestion, phases of the 31
process of Gegenbaur, of the fourth lumbar vertebra 23
process of liver, caudate 37
process of liver, papillary 37
process of malleus 64
process of malleus, anterior 66
process of malleus, lateral 66
process of mandible coronoid 17
process of maxilla, alveolar 17
process of scapula coracoid 14, 19
process of talus, posterior 18
process of the fifth lumbar vertebra, spinous 23
process of the fifth lumbar vertebra, transverse articular 23
process of the fifth thoracic vertebra, costal facet of transverse 23
process of the fifth thoracic vertebra, inferior articular 22
process of the fifth thoracic vertebra, spinous 22
process of the fifth thoracic vertebra, superior articular 22
process of the fifth thoracic vertebra, transverse 22
process of the fourth cervical vertebra, hairpin of bifid spinous 23
process of the fourth cervical vertebra, transverse 23
process of the fourth cervical vertebra, unciform 23
process of the fourth lumbar vertebra, inferior articular 23
process of the fourth lumbar vertebra, spinous 23
process of the fourth lumbar vertebra, superior articular 23
process of the sacrum bone, superior articular 22
process of the seventh cervical vertebra, transverse 23
process of the sixth thoracic vertebra, spinous 22
process of the sixth thoracic vertebra, superior articular 22
process of the sixth thoracic vertebra, transverse 22
process, accesory of fourth lumbar vertebra 23
process, annular of the fifth thoracic vertebra 22
process, ciliary 62
process, costal of the fourth lumbar vertebra 23
process, development of auditory 66
process, forked spinous of the fourth cervical vertebra 23
process, inferior articular of the fifth lumbar vertebra 23
process, inferior articular of the sixth thoracic vertebra 22
process, lingual 68
process, mamillary of the fourth lumbar vertebra 23
process, mastoid 17, 66
process, maxillary frontal 14, 61, 67
process, semilunar of the seventh cervical vertebra 23
process, spinous of the seventh cervical vertebra 23
process, styloid 17, 18, 19, 66
process, thoracolumbar 25
process, ulnar 19
process, xiphoid 14
projection of bronchii on lung wall 43
projection of colon on anterior abdominal wall 39
projection of duodenum on anterior abdominal wall 34
projection of heart on anterior abdominal wall 48
projection of large intestine on anterior abdominal wall 39
projection of liver on anterior abdominal wall 36
projection of pancreas on anterior abdominal wall 34
projection of small intestine on anterior abdominal wall 38
projection of stomach on anterior abdominal wall 32
prominence of epitympanic process 64
prominence of epitympanic recess 64
promontory of sacrum 21, 22, 88
promontory of tympanic cavity 66
promontory of tympanic cavity, sulcus of 66
promontory of tympanic membrane 64
pronator quadratus muscle 28
pronator teres muscle 24, 26
prostate 77, 78
prostate, posterior surface of 77
prostatic plexus 78
protein 31
protuberance, external occipital 15, 17
psoas major muscle 24, 35
pubic bone 6, 7, 14
pubic hair 80
pubic region 8
pubic tubercle 16
pubis, mons 79
pudendal nerve 55
pulmonary artery 46, 47
pulmonary artery, bifurcation of trunk of 49
pulmonary artery, left 48
pulmonary artery, right 49
pulmonary artery, valves of 48
pulmonary segments, apical 44, 45
pulmonary segments, inferior of lingula 44, 45
pulmonary segments, lateral 44, 45
pulmonary segments, lateral basal 44, 45
pulmonary segments, lung 44, 45
pulmonary segments, lung basal 44, 45
pulmonary segments, medial 44, 45
pulmonary segments, medial basal 44, 45
pulmonary segments, posterior 44, 45
pulmonary segments, posterior basal 44, 45
pulmonary segments, superior 44, 45
pulmonary segments, superior of lingula 44, 45
pulmonary vein 46, 47
pulmonary vein, left 48, 49
pulmonary vein, right 48, 49
pulp of incisor tooth 41
pulp of premolar tooth 42
puncta, lacrimal 61
pupil 59, 61, 62
pupil, dilator muscle of 59
pupil, margin of 62
pupil, sphincter of 59
pyloric antrum of stomach 32
pyloric canal 34
pyloric sphincter muscle 34
pylorus 32, 33
pyramid of brain 54
pyramid of Malpighi 75
pyramidal decussation of brain 54
pyramidalis nasi muscle 24, 27

Q

quadratus lumborum muscle 35

R

radial artery 47
radial nerve 55, 58
radial nerve, dorsal digital branches of superficial branch of 55
radiation of Gatriolet, optic 62
radius 13, 14, 15, 16, 18, 19, 26, 28
ramification of a triple root tooth, arterial 41
ramification of a triple root tooth, nervous 41
ramification of a triple root tooth, venous 41
ramus communicantes of spinal cord 78, 79
ramus of mandible 17
ramus of spinal nerve, anterior cutaneous branch 57
ramus of spinal nerve, cutaneous branch 57
ramus of spinal nerve, deep branch 57
ramus of spinal nerve, dorsal (afferent) 57
ramus of spinal nerve, gray linking ramus communicans 57
ramus of spinal nerve, lateral cutaneous branch 57
ramus of spinal nerve, meningeal branch 57
ramus of spinal nerve, muscular branch 57
ramus of spinal nerve, superficial branch 57
ramus of spinal nerve, white ramus communicans 57
raphe muscle 25
rays, optic 62, 63
recess, bottom of rectovesical 78
recess, epitympanic 64
recess, rectovesical 78
rectum 31, 79, 88
rectum, ampulla of 78
rectus abdominis muscle 24
rectus abdominis muscle, sheath of 26
rectus femoris muscle 24, 26
rectus muscle, inferior 59, 61
rectus muscle, lateral 60, 61
rectus muscle, medial 60
rectus muscle, superior 61
reflex act 73
region of axillary cavity 8
region of axillary fossa 8
region of infraclavicular fossa 8
region of palm of hand 8, 9
region of popliteal fossa 9
region of the elbow, posterior 9
region of the knee 8
region of the knee, posterior 9
region of the leg, posterior 8, 9
region of the neck, lateral 8
region of the neck, posterior 9
region of thorax, lateral 8, 9
region of triangle of Scarpa (femoral triangle) 8
region, anal 9
region, antebrachial 9
region, anterior antebrachial 8, 9
region, anterior brachial 9
region, anterior femoral 8
region, anterior of the elbow 8
region, anterior of the knee 8
region, anterior of the leg 8, 9
region, anterior of the neck 8
region, axillary 9
region, brachial 9
region, buccal 9
region, calcaneal 8, 9
region, cubital 8, 9
region, deltoid 8, 9
region, dorsum of foot 8, 9
region, dorsum of hand 8
region, epigastric 8
region, femoral 8
region, gluteal 9
region, hypochondrial 8
region, inframammary 8
region, intrascapular 9
region, inguinal 8
region, internal malleolar 8
region, interscapular 9
region, labial 8
region, left lumbar 8, 9
region, lumbar 9
region, mammary 8
region, mental 8
region, nasal 8
region, occipital 9
region, orbital 8, 9
region, parietal 8, 9
region, parotidmasseteric 9
region, pectoral 8
region, plantar 9
region, posterior antebrachial 8, 9
region, posterior brachial 9
region, posterior femoral 8
region, presternal 8
region, pubic 8
region, sacral 9
region, scapular 9
region, sternocleidomastoideal 8, 9
region, subclavian 8
region, submandibular 9
region, suprascapular 9
region, temporal 8, 9
region, trochanteric 8
region, umbilical 8
region, urogenital 8
region, vertebral 9

region, zygomatical 9
renal artery 35, 75
renal vein 35
representation of the DNA
    molecule, schematic 11
reproductive organs, autonomic
    innervation of female 79
reproductive organs, autonomic
    innervation of male 78
reproductive organs, female 79
reproductive organs, male 78
reproductive system 78–90
residue 31
respiratory system 43–46
response of the organism to
    external aggressions 73
resting forearm 30
rete testis 80
reticulum, rough endoplasmic 10
reticulum, sarcoplasmic 29
reticulum, smooth endoplasmic 10
reticulum, trabecular 62
retina 59, 60, 63
retina, blood vessel of 59
retina, capillary bed of 59
retina, central artery of 59
retina, central vein of 59
retina, cerebral stratum of 59
retina, ciliary area of 59
retina, cone of 59
retina, external limiting membrane
    of 59
retina, internal limiting membrane
    of 59
retina, layer of cones and rods of 59
retina, neuroepithelial stratum of 59
retina, nuclear external layer of 59
retina, nuclear internal layer of 59
retina, pigment cell of 59
retina, pigmented epithelium of 59
retina, pigmented stratum of 59
retina, rod of 59
retina, section of 59
retina, vitreous body of 59
retinaculum, extensor 25, 26
retinaculum, flexor 24, 25
retinaculum, inferior extensor 26
retinaculum, inferior peroneal 25, 28
retinaculum, peroneal 25, 26
retinaculum, superior extensor 24
retinaculum, superior
    peroneal 25, 28
rhomboideus major muscle 25
rib 13, 14, 15, 90
rib, eighth 16
rib, eleventh 16
rib, fifth 16
rib, first 16
rib, fourth 16
rib, ninth 16
rib, second 16
rib, seventh 16
rib, sixth 16
rib, tenth 16
rib, third 16
rib, twelfth 16
ribosome 10
ring finger 18
ring, left fibrous 48
ring, of iris major 62
ring, of iris minor 62
ring, of spermatozoon terminal 80
ring, right fibrous 48
ring, tympanic 64
risorius muscle (of Santorini) 24,
    26, 27
rod of retina 59
rods of retina, layer of cones and 59
Rolando, sulcus of 53
root canal of a premolar tooth 42
root canal of a premolar tooth,
    foramen of 42

root canal of an incisor tooth 41
root canal of tooth 41
root canal of tooth, foramen of 41
root of a premolar tooth 42
root of a spinal nerve, efferent 57
root of antihelix 64
root of brain, lateral 63
root of brain, middle 63
root of helix 64
root of nail 73
root of spinal cord, ventral
    (efferent) 79
root of spinal nerve, ventral
    (efferent) 57
root of teeth, imprint on mandible
    of 41
root of tongue 68, 69
root of tooth 41
root, section of mandible with
    teeth of triple 41
root, tooth of double 41
root, tooth of single 41
root, tooth of triple 41
roots of I spinal nerve, ventral 54
roundlet of nail 73
Ruffini's, corpuscle 56, 71

S

sac, lacrimal 61
sac, viteline 87
saccule of osseous labyrinth
    of ear 65
sacral nerve 58, 79
sacral region 9
sacrum bone 13, 14, 15, 16, 21, 57,
    58, 78, 79
sacrum bone, anterior foramina 22
sacrum bone, anterior view of 22
sacrum bone, apex of 22
sacrum bone, foramen of 22
sacrum bone, lateral area of 22
sacrum bone, lateral view of 22
sacrum bone, plexus of 79
sacrum bone, promontory of 21,
    22, 88
sacrum bone, transverse lines of 22
sacrum bone, tuberosity of 22
sacrum bone, wing of 22
sacrum, auricular surface of 22
sacrum, base of 22
sacrum, medium crest of 22
sacrum, superior articular facet
    of 22
saliva 31, 70
salt 31
Santorini, duct of 34
Santorini, risorius muscle of 24
saphenous nerve 55, 58
saphenous nerve, internal 55, 58
saphenous vein, internal 47
saphenous vein, lateral 47
sarcolemma 29
sarcomere 29
sarcoplasmic reticulum 29
sartorius muscle 24, 25, 26
sartorius tendon 24
scala tympani 66
scala tympani of osseous cochlea 65
scala vestibuli of ear 66
scala vestibuli of osseous cochlea 65
scalp 72
scaphoid bone 14, 18
scapula 6, 7, 13, 14, 15, 16, 19
scapula, angle of 15
scapula, dorsal nerve of 55
scapula, inferior angle of 19
scapula, lateral angle of 19
scapula, spine of 16, 25, 26
scapula, superior angle of 19
scapular region 9
Scarpa, region of triangle (femoral
    triangle) of 8

schematic longitudinal section of a
    premolar tooth 42
schematic section of a blood
    capillary 52
schematic section of a finger 73
schematic section of a hair 71
schematic section of a main
    artery 52
schematic section of a main
    vein 52
schematic section of a vein 52
schematic section of an artery 52
schematic section of an ovary 81
schematic section of different
    types of skin of the human
    body 72
schematic section of left kidney 75
schematic section of mandible with
    triple root teeth 41
schematic section of olfactory
    mucosa 67
schematic section of the skin 71
schematic section of wall of
    colon 39
schematic section of wall of
    ileum 38
schematic section of wall of
    stomach 33
schematic view of a vein 52
Schwalde, line of 59
Schwann cell of a motor neuron 56
sciatic nerve 55
sciatic nerve, great, 55
sciaticnerve, lesser 55
sclera 59, 60
sclera, sinus venosus of 59
scrotum 6, 78, 80
sebaceous gland 72
section of a hair 72
section of an eye, horizontal, 59
section of an incisor tooth,
    longitudinal 41
section of ascending colon,
    coronal 39
section of cecum, coronal 39
section of retina 59
segment identified by colors,
    anterior view of
    bronchopulmonary 44
segment, pelvic 21
sella turcica 67
semilunar cusp of aortic valve
    of posterior 48
semimembranous muscle 25, 26
semispinalis capitis muscle 27
semitendinous muscle 25
sensation of pain 73
sensation of taste, transmission
    of a 70
sense 59, 73
sensory receptor for cold 71
sensory receptor for heat 71
sensory receptor for pain 71
sensory receptor for pressure and
    vibration 71
sensory receptor for touch 71
sensory receptor of a unipolar
    neuron 56
septi nasi muscle, depressor 27
septum of placenta 86
septum of testis 80
septum, anterior intermuscular 26
septum, cartilage of nasal 67
septum, interalveolar 46
septum, lingual 68
septum, medial intermuscular 26
septum, membranous part of
    interventricular 48
septum, muscular part of
    interventricular 48
septum, nasal 17

septum, orbital 61
septum, rectovaginal 79, 88
serratus anterior muscle 26, 90
serratus magnus muscle 24
serratus nerve 55
sesamoid bone 18
seventh cervical vertebra 23
seventh cervical vertebra, apex of
    spinous process of 23
seventh cervical vertebra, lamina
    of 23
seventh cervical vertebra,
    peduncle of vertebral arch of 23
seventh cervical vertebra,
    semilunar process of 23
seventh cervical vertebra, spinous
    process of 23
seventh cervical vertebra, superior
    articular process of 23
seventh cervical vertebra,
    transverse foramen of 23
seventh cervical vertebra,
    transverse process of 23
seventh cervical vertebra,
    vertebral arch of 23
seventh cervical vertebra,
    vertebral body of 23
sheath of fingers, annular part of
    fibrous 28
sheath of fingers, fibrous 24
sheath of flexor pollicis longus
    tendon 28
sheath of hair, external root 72
sheath of myelin of a motor
    neuron 56
sheath of myelin of a multipolar
    neuron 56
sheath of optic nerve 60
sheath of peroneus tendons,
    common 28
sheath of spermatozoion,
    mitochondrial 80
sheath, cruciate part of fibrous 28
sheath, occipital 17
sheath, rectus 26
shin 6, 7
short head of biceps femoris
    muscle 26
short muscle 29
shoulder 6, 7
Shrapell, membrane of 64
sight 59, 60, 61, 62, 63
sight, main defects and their
    correction 63
Silvio, fissure of (lateral
    sulcus of) 53
sinus of prostatic area of male
    urethra, prostatic 77
sinus of sclera, venous 59
sinus of tarsus 18
sinus of tympanic cavity,
    posterior 66
sinus of vena cava 49
sinus, frontal 67
sinus, maxillary 61
sinus, renal 75
sinus, sphenoidal 67
sinus, transverse pericardial 49
sixth thoracic vertebra 22
sixth thoracic vertebra, inferior
    articular process of 22
sixth thoracic vertebra, inferior
    costal facet of 22
sixth thoracic vertebra, inferior
    vertebral notch of 22
sixth thoracic vertebra, spinous
    process of 22
sixth thoracic vertebra, superior
    articular facet of the vertebral
    body of 22
sixth thoracic vertebra, superior
    articular process of 22

sixth thoracic vertebra, superior
    costal facet of 22
sixth thoracic vertebra, transverse
    process of 22
sixth thoracic vertebra, vertebral
    body of 22
skeleton 14, 15, 16
skeleton, anterior view 14
skeleton, lateral view 16
skeleton, posterior view 15
skin 61, 72, 73
skin, adipose tissue of 71
skin, fat of 71
skin, papilla of 71
skin, schematic section of 71
skin, stratum basale of 71
skin, stratum corneum of 71
skin, stratum granulosum of 71
skin, stratum spinosum of 71
skull 17
skull, anterior view of the 17
skull, lateral view of the 17
skull, posterior view of the 17
skull, superior view of the 17
sole of foot 72
soleus muscle 24, 25, 26
space of Bowman 76
space, cochlear perilymphatic
    (of ear) 66
space, perichordial 59
space, perichoroid 60
space, retropubic 78
space, suprasternal 6, 7
spermatid 80
spermatogenesis 80
spermatogonia 80
spermatozoon 80, 83, 85
spermatozoon and oocyte,
    meeting of the 84
spermatozoon, (genetic material)
    phases until its implantation in
    the uterus 83
spermatozoon, flagellum of the 80
spermatozoon, fusion of oocyte
    with the 80
spermatozoon, head of the 80
spermatozoon, meeting of oocyte
    with the 85
spermatozoon, middle portion
    of the 80
spermatozoon, mitochondrial
    sheath of the 80
spermatozoon, nucleus of the 80
spermatozoon, primary 80
spermatozoon, secondary 80
spermatozoon, tail of the 80
spermatozoon, terminal ring
    of the 80
sphenoid bone 14, 16, 17, 61, 67
sphenoid papilla 70
sphincter muscle, pyloric 31, 34
spinal cord 54, 57, 73, 78, 79, 88
spinal cord, central duct of 73
spinal cord, cutaneous areas
    innervated by fragment of 57
spinal cord, dorsal cord of 57
spinal cord, dorsal horn of 57, 73
spinal cord, dorsal median sulcus
    of 73
spinal cord, dorsal root of 79
spinal cord, interganglionar
    branches of 79
spinal cord, intermediate
    substance of 57
spinal cord, lateral cord of 57
spinal cord, lateral white matter 57
spinal cord, posterior white
    matter 57
spinal cord, ramus communicantes
    of 78, 79
spinal cord, spinal nerves 57, 58

spinal cord, substantia gelatinosa
    of 73
spinal cord, ventral horn of 57, 73
spinal cord, ventral root of 79
spinal cord, white matter of 73
spinal nerve 58
spinal nerve, ventral roots of I 54
spinal nerves, gray matter 57
spinal nerves, white matter 57
spindle, muscule 73
spine of scapula 15, 16, 25, 26
spine, anterior inferior iliac 14, 20
spine, anterior nasal 17
spine, anterior superior iliac 14, 24
spine, iliac 20
spine, ischial 16
spine, posterior inferior iliac 15, 16
spine, posterior superior iliac 15
spine, sciatic 15
spinous process 16, 21, 22, 23, 88
spiral organ (of Corti) 65
splanchnic nerve, pelvic 79
spleen 31, 40
spleen, anterior extremity of 40
spleen, anteromedial view of 40
spleen, colic surface of 40
spleen, craneal lateral view of 40
spleen, diaphragmatic surface of 40
spleen, fibrous layer of 40
spleen, gastric surface of 40
spleen, hilus of 40
spleen, inferior margin of 40
spleen, parenchymal extremity of 40
spleen, posterior extremity of 40
spleen, renal surface of 40
spleen, schematic caudal view of 40
spleen, superior margin of 40
spleen, trabeculae of 40
spleen, visceral surface of 40
splenic artery 40, 47
splenic artery, sulcus of 35
splenic vein 35, 40
splenium of corpus callosum 63
splenius muscle 26
splenius muscle of the head 25, 27
spongiosum, urethral corpus 78
stapes 12
stapes bone 65, 66
stapes, anterior crus of 66
stapes, base of 66
stapes, long crus of 64
stapes, posterior crus of 66
stay of free egg in the uterine
    cavity 84
sternocleidomastoid muscle 24, 25,
    26, 27
sternocleidomastoideal region 8, 9
sternum 13, 14, 16, 88
stomach 31, 32, 33, 88
stomach cupola of 32
stomach, angular notch of 32
stomach, anterior view of the 32
stomach, anterior view of the 33
stomach, body of the 32, 33
stomach, cardiac notch of 32
stomach, cardiac zone of the 33
stomach, circular muscle layer of 33
stomach, fundus of 32, 33
stomach, gastric area of tunica
    mucosa of the 33
stomach, gastric folds of the 32
stomach, gastric fossae of mucosa
    of the 33
stomach, general view of the 32
stomach, gland (gastric) of
    mucosa of 33
stomach, glands of mucosa
    of the 31
stomach, greater curvature
    of 32, 33
stomach, imprint of 37

stomach, lesser curvature of 32, 33
stomach, longitudinal muscle layer of 33
stomach, microscopic view of the interior wall of the 32
stomach, mucosa of the wall of the 33
stomach, mucosal epithelium of the 33
stomach, muscularis externus 33
stomach, muscularis mucosa 33
stomach, projection on abdominal wall of 32
stomach, pyloric antrum of 32
stomach, pyloric part of the 32
stomach, schematic section of the wall of the 33
stomach, serosa of the 33
stomach, serosal layer 33
stomach, solitary lymph lamina propria of the 33
stomach, submucosa of the 33
stratum corneum of skin 71
stratum granulosum of skin 71
stratum of retina, cerebral 59
stratum of retina, neuroepithelial 59
stratum of retina, pigmented 59
stratum spinosum of skin 71
stratum, basale of skin 71
structure of a motor neuron 56
structure of an ovum 81
structure of contact of epidermis 71
subclavian artery 47
subclavian region 8
subclavian vein 47
submandibular region 9
substance of brain, anterior perforated 54
substance of spinal cord, intermediate 57
substance, toxic 31
substantia adamantina (enamel) of teeth 41
substantia gelatinosa of spinal cord 73
substantia propria of cornea 59
sugar/phosphate, link formed by 11
sulcus of brain, central 53
sulcus of brain, cingulate 53
sulcus of brain, inferior frontal 53
sulcus of brain, interior temporal 54
sulcus of brain, intraparietal 53
sulcus of brain, lateral 53
sulcus of brain, parieto-occipital 53
sulcus of brain, postcentral 53
sulcus of brain, precentral 53
sulcus of brain, superior frontal 53
sulcus of brain, superior temporal 53
sulcus of brian, orbital 53
sulcus of Rolando (central) 53
sulcus of Silvio (lateral) 53
sulcus of spinal cord, dorsal median 73
sulcus of splenic artery 35
sulcus of the promontory of tympanic cavity 66
sulcus of tongue, middle 68, 69
sulcus of tongue, terminal 68
sulcus of vertebral artery of atlas 23
sulcus, calcarine 62, 63
sulcus, coronary 49
sulcus, dorsal median of brain 54
sulcus, dorsal median of vertebra 77
sulcus, malleolar 20
sulcus, olfactory 54
sulcus, superior palpebral 61
sulcus, terminal 49
sulcus, tympanic 66
superior bronchial segment 44, 45
superior colliculus 62
superior colliculus of brain, arm of 63
superior gluteal nerve 55
supinator muscle 24
support cell of olfactory membrane 67
support cells of olfactory mucosa 67
supraorbital nerve 42
suprarenal artery, inferior 75
suprarenal artery, middle 75
suprarenal artery, superior 75
suprarenal gland (adrenal) 35, 75
suprarenal vein 75
suprascapular nerve 55
suprascapular region 9
sural nerve 58
surface joint of femur toward patella 20
surface of acromial joint of acromion 19
surface of acromial joint of clavicle 19
surface of acromial joint of tibia toward fibula 20
surface of cornea, posterior 62
surface of fibula, lateral 20
surface of iliac bone, auricular 20
surface of inferior articular joint of tibia 20
surface of iris, posterior 62
surface of left lobe of liver, diaphragmatic 36
surface of liver, diaphragmatic 37
surface of lung, diaphragmatic 43
surface of malleolar articular joint of tibia 20
surface of olfactory mucosa 67
surface of os coxae bone, auricular 20
surface of patella, articular 20
surface of posterior arch of atlas, ventral 23
surface of prostate, posterior 77
surface of right lobe of liver, diaphragmatic 36
surface of sacrum bone, auricular 20, 22
surface of spleen, diaphragmatic 40
surface of spleen, gastric 40
surface of spleen, renal 40
surface of spleen, visceral (colic) 40
surface of sternal joint of clavicle 19
surface of superior articular joint of tibia 20
surface of superior lobe of left lung, lateral 43
surface of superior lobe of right lung, lateral 43
surface of talus lateral malleolar 18
surface of the vertebral body of axis, inferior articular 22
surface of the vertebral body of fourth cervical vertebra, superior articular 22
surface of the vertebral body of sixth thoracic vertebra, superior articular 22
surface of tibia, lateral 20
surface of tibia, medial 20
surface of tongue, inferior 68
surface of ulna, posterior 19
surface of vertebral body of fifth thoracic vertebra intervertebral 22
surface of vertebral body of fourth lumbar vertebra intervertebral 23
surface, anterior patellar 20
surface, anterior radial 19
surface, anterior ulnar 19
surface, anterolateral of humerus 19
surface, anteromedial of humerus 19
suspensory muscle of duodenum 34, 35
suture, coronal 17
suture, frontomaxillary 67
suture, frontonasal 67
suture, frontoparietal 14, 16
suture, internasal 14
suture, interparietal 15
suture, lamboid 15, 17
suture, nasomaxillary 17, 67
suture, occipitomastoid 17
suture, parietomastoid 17
suture, parietooccipital 16
suture, saggital 15, 17
suture, sphenosquamous 17
suture, squamous 16, 17
suture, temporal zygomatic 17
symphysis 14, 25, 78, 79, 88
symphysis pubis 14, 25, 78, 79, 88
system of fetus, digestive 87
system, circulatory 47–52
system, main veins and arteries of the circulatory 47
system, muscular 24–30
system, nervous 53–58, 87
system, peripheral nervous 55
system, skeletal 12–23
systole, atrial 50
systole, ventricular 50

T

tail 87
tail of epididymis 74, 78, 80
tail of pancreas 34, 35
tail of spermatozoon 80
talus (ankle bone) 64
talus, bone 14, 15, 16, 18
talus, head of 18
talus, lateral malleolar surface of 18
talus, posterior process of 18
talus, trochlea of 18
tarsus 18, 20
tarsus, anterior annular ligament of 28
tarsus, bones of 13
tarsus, sinus of 18
taste 68, 69, 70
taste bud 70
taste bud of a circumvallate papilla 69
taste bud, cross section 69
taste bud, taste cell of 69
taste, area of perception of 69
taste, area of perception of acid 69
taste, area of perception of salty 69
taste, area of perception of sour 69
taste, primary 70
teeth, deciduous 41
teeth, permanent 42
teeth, types of 41
temperature, basal 84
temporal bone 12, 14, 15, 16, 64
temporal muscle 26
temporal region 8, 9
temporalis muscle 27
tendon 29
tendon of Achilles 25, 26, 28
tendon of biceps brachii 24
tendon of extensor digitorum longus muscle 26
tendon of extensor digitorum muscle 26
tendon of tensor tympani muscle 64
tendon of triceps brachii muscle 25, 26
tendon, bipenniform 29
tendon, common sheath of peroneus 28
tendon, extensor digitorum longus 28
tendon, extensor hallucis 28
tendon, extensor hallucis longus 28
tendon, extensor pollicis brevis 28
tendon, flexor digitorum profundus 28
tendon, gastrocnemius 25
tendon, intermediate bipenniform 29
tendon, peroneus brevis 25, 26, 28
tendon, peroneus longus 24, 25, 28
tendon, peroneus tertius 26
tendon, plantaris 25
tendon, sartorius 24
tendon, sheath of flexor pollicis longus 28
tendon, tibialis anterior 26, 28
tenia, coli 31, 39
tensor fasciae latae muscle 24
tensor tympani muscle 64
tensor tympani tendon 64
teres major muscle 25, 26
teres minor muscle 26
testicular artery 35, 78
testicular plexus 78
testicular vein 35, 75
testis 74, 78, 80
testis, apex of 74, 78, 80
testis, defferential plexus of 78
testis, ductus deferens of 78, 80
testis, lobe of 78, 80
testis, rete 80
testis, seminiferous tubles of 80
testis, septum of 80
thalamus 70
thigh 6, 7
thigh, lateral femoral cutaneous nerve of 56, 58
thigh, posterior femoral cutaneous nerve of 58
thigh, rectus femoris muscle of 34
thoracic nerve 58
thoracic or dorsal vertebra 21
thorax 6, 7
thorax of fetus 87
thorax, lateral region of 8, 9
thumb 6, 7
thumb, 18
thumb, abductor pollicis longus muscle of 24
thumb, abductor pollicis muscle of 28
thumb, extensor brevis tendon of 28
thumb, extensor pollicis brevis muscle of 25, 26
thumb, extensor pollicis longus tendon of 28
thumb, extensor pollicis tendon of 28
thumb, flexor longus pollicis of 25
thumb, flexor pollicis longus muscle 24, 26
thumb, flexor pollicis muscle of 28
thumb, opponens pollicis longus muscle of 26
thumb, opponens pollicis muscle of 28
thumb, sheath of flexor pollicis longus tendon of 28
thymine 11
thyroid cartilage 68
tibia bone 13, 14, 15, 16, 20, 28
tibia, anterior margin of 16
tibial artery, anterior 47
tibial artery, posterior 47
tibial nerve 55
tibial nerve, plantar branches of 58
tibial tuberosity 26
tibial vein 47
tibial vein, posterior 47
tibialis anterior muscle 24, 26, 28
tibialis posterior tendon 26, 28
time of eruption of teeth 41, 42
tip of a premolar tooth 42
tissue of a hair, conjunctival 72
tissue of alveolus, connective 46
tissue of an artery, soft 52
tissue of an incisor tooth, conjunctival 41
tissue of a premolar tooth, conjunctival 42
tissue of skin, adipose 71
tissue, lymphatic 31
toe 18
toe, base of phalanx of 18
toe, body of phalanx of 18
toe, distal phalanx of 18
toe, extensor hallucis brevis muscle of 26, 28
toe, extensor hallucis longus muscle of 24
toe, extensor hallucis longus tendon of 28
toe, extensor hallucis tendon of 28
toe, head of phalanx of 18
toe, middle phalanx of 18
toe, proximal phalanx of 18
toes, extensor digitorum longus muscle of 28
tongue 31, 70
tongue fragment in secton 69
tongue, apex of 68
tongue, back of 68, 69
tongue, body of 68
tongue, circumvallate papilla of 68
tongue, dorsal view of 68
tongue, filiform papilla of 68
tongue, folicate papillae of 68
tongue, foramen cecum of 68
tongue, fungiform papillae of 68
tongue, inferior longitudinal muscle of 68
tongue, inferior surface of 68
tongue, lateral glossoepiglottic fold of 68
tongue, lateral view of 68
tongue, margin of 68
tongue, median sulcus of 68, 69
tongue, pons of 68
tongue, postsulcal area of 68
tongue, presulcal area of 68
tongue, root of 68, 69
tongue, schematic section of part of the 69
tongue, sublingual fold of 68
tongue, sulcus terminalis of 68
tongue, superior longitudinal muscle of 68
tongue, superior view of 69
tongue, transverse muscle of 68
tongue, transverse section of 68
tongue, triangular fold of 68
tongue, vertical muscle of 68
tonsil, adenoid 67
tonsil, lingual 68
tonsil, palatine 68
tonsil, pharyngeal 67
tooth 31, 41, 67, 68
tooth with double root 41
tooth with one root 41
tooth with triple root 41
tooth with triple root, artery of a 41
tooth with triple root, cavity of a 41
tooth with triple root, central cavity of a 41
tooth with triple root, crown of a 41
tooth with triple root, enamel of a 41
tooth with triple root, neck of a 41
tooth with triple root, nerve of a 41
tooth with triple root, root canal of a 41
tooth with triple root, section of mandible with a 41
tooth with triple root, vein of a 41
tooth, age of eruption of a 41, 42
tooth, bone of a 41
tooth, canine 41, 42
tooth, cementum of a premolar 42
tooth, central cavity of a premolar 42
tooth, conjunctival tissue of a premolar 42
tooth, conjunctival tissue of an incisor 41
tooth, crown of a premolar 42
tooth, enamel (substantia adamantina) of an incisor 41
tooth, enamel of a premolar 42
tooth, gingival (gum) epithelium of an incisor 41
tooth, gingival epithelium of a premolar 42
tooth, imprint of teeth roots in mandible bone 41, 42
tooth, imprint of teeth roots in maxilla bone of a 41, 42
tooth, incisor 41, 42
tooth, innervation of 42
tooth, location in mouth of a 42
tooth, longitudinal section of a premolar 42
tooth, longitudinal section of an incisor 41
tooth, lower 41, 42
tooth, molar 41, 42
tooth, neck of a premolar 42
tooth, premolar 41, 42
tooth, pulp of a premolar 42
tooth, pulp or central cavity of an incisor 41
tooth, root canal of a premolar 42
tooth, root foramen of an incisor 41
tooth, root of a 41, 42
tooth, root of a premolar 42
tooth, superior 41, 42
tooth, superior view of a 41
tooth, tip of a premolar 42
top of midbrain 63
torus tubarius 67
touch 71, 72, 73
trabeculae of corpus cavernosus of male urethra 77
trabeculae of corpus cavernosus of penis 78
trabeculae of spleen 40
trabeculae, muscular 48
trachea 31, 44, 45
trachea, annular ligament of 43, 45
trachea, anterior view of 44, 45
trachea, cartilage of 43
trachea, membranous part of 45
tracheal bifurcation (carina) 43, 44
tract, iliotibial 26
tract, olfactory 54, 63, 67
tract, optic 62, 63
tragus 64
transmission of a taste sensation 70
transversalis cervicis muscle 25
transverse (cervical) nerve 55, 58
transverse muscle of the tongue 68
transverse nerve of neck 55, 58
transverse section of tongue 68
transverse section of upper part of small intestine 38
trapezium bone 14, 18
trapezius muscle 24, 25, 26, 27
trapezoid bone 14, 18
Treitz, (suspensory) muscle of 34
triangle of neck, muscle 24, 25
triangle of Scarpa, region of 8
triceps brachii muscle 24, 25
triceps brachii muscle, long head of 26
triceps brachii muscle, medial head of 26

triceps brachii muscle, short head of 26
triceps brachii tendon 26
triceps in resting phase 30
triceps, completely relaxed 30
triceps, contracted 30
triceps, relaxed 30
trigeminal nerve 42, 54
trigeminal nerve, ganglion of 42, 54
trigeminal nerve, motor root of 54
trigone of myocardium, left fibrous 48
trigone of myocardium, right fibrous 48
trigone of urinary bladder 77, 78
trigone, left fibrous 48
trigone, right fibrous 48
triquetral bone 14, 18
trochanter, greater 14, 20
trochanter, lesser 14, 20
trochanteric region 8
trochlea 18, 61
trochlea of calcaneus, peroneal 18
trochlea of talus 18
trochlea, humeral 14, 19
trochlear nerve 54, 61
trophoblast 86
trunk and limbs, bones of 13
trunk, bifurcation of pulmonary 49
trunk, celiac 35
trunk, ganglia of sympathetic 78, 79
trunk, pulmonary 49
trunk, valves of pulmonary 48
tube, abdominal foramen of the Fallopian 79
tube, abdominal foramen of the uterine (or Fallopian) 79
tube, ampulla of the uterine (or Fallopian) 79
tube, descent of ovum through the uterine (or Fallopian) 84
tube, Eustachian 65
tube, Fallopian 79, 81, 82, 83, 85
tube, infundibula of the uterine (or Fallopian) 74, 79, 82, 83
tube, isthmus of the uterine (or Fallopian) 79
tube, opening of auditory (Eustachian) 67
tube, pathway of oocyte to uterus through the uterine (or Fallopian) 82
tube, tympanic orifice of auditory 66
tube, uterine (or Fallopian) 79, 81, 82, 83, 85
tuber cinereum 54, 63
tubercle of atlas, anterior 23
tubercle of atlas, posterior 23
tubercle of brain, mammillary 54
tubercle of clavicle, conoid 19
tubercle of femur 20
tubercle of fifth lumbar vertebra, mammillary 23
tubercle of fifth metatarsal bone 18
tubercle of humerus, greater 19
tubercle of humerus, lesser 19
tubercle of liver, omental 37
tubercle of metatarsal bone 18
tubercle of pubis 16, 20
tubercle, greater 14, 15
tubercle, lesser 14
tuberosity of atlas for insertion of transverse ligaments of axis 23
tuberosity of calcaneous 18, 25, 26
tuberosity of humerus, deltoid 19
tuberosity of ischium 15, 16, 20
tuberosity of phalanx of finger 18
tuberosity of radius 19
tuberosity of sacrum bone 22
tuberosity of tibia 14, 20, 26
tuberosity of ulna 19

tuberosity, calcaneal 25, 26
tuberosity, iliac 20
tuberosity, ischial 15, 20
tubule of a nephron, collecting 76
tubule of testis, seminiferous 80
tunica adventitia of a main artery 52
tunica adventitia of a main vein 52
tunica adventitia of an artery 52
tunica albuginea 80
tunica albuginea of corpus cavernosus of male urethra 77
tunica albuginea of corpus cavernosus of penis 78
tunica intima of a main artery 52
tunica intima of a main vein 52
tunica intima of an artery 52
tunica media, artery 52
tunica media, vein 52
tunica mucosa of gallbladder 37
tunica mucosa of upper part of small intestine, intestinal villi of 38
tunica mucosa of upper part of small intestine, muscularis mucosa of 38
tunica mucosa of urinary bladder 77
tunica mucosa of wall of colon, epithelium of 39
tunica mucosa of wall of colon, intestinal gland of 39
tunica mucosa of wall of colon, lamina propria of 39
tunica mucosa of wall of colon, lymphatic follicle of 39
tunica mucosa of wall of colon, muscularis mucosa of 39
tunica mucosa of wall of illeum, central lymph vessel of 38
tunica mucosa of wall of illeum, intestinal gland of 38
tunica mucosa of wall of illeum, lamina propria of 38
tunica mucosa of wall of illeum, muscularis mucosa of 38
tunica mucosa of wall of stomach 33
tunica mucosa of wall of stomach, epithelium of 33
tunica mucosa of wall of stomach, gastric area of 33
tunica mucosa of wall of stomach, gastric fossae of 33
tunica mucosa of wall of stomach, gastric gland of 33
tunica mucosa of wall of stomach, muscularis mucosa of 33
tunica mucosa of wall of stomach, solitary lymphatic follicle of lamina propria of 33
tunica serosa 79
tunica serosa of peritoneum of small intestine 38
tunica serosa of spleen 40
tunica serosa of wall of colon 39
tunica serosa of wall of gallbladder 37
tunica serosa of wall of illeum 38
tunica serosa of wall of stomach 33
turbinated bone 12
tympani, chorda 64
tympanus 65
tympanus, cavity of 66
tympanus, facial duct of the cavity of 66
tympanus, mastoid antrum of cavity of 66
tympanus, mastoid wall of cavity of 66
tympanus, membrane of 64, 66
tympanus, posterior sinus of the cavity of 66
tympanus, promontory of the cavity of 66

tympanus, sulcus of promontory of the cavity of 66
tympanus, tendon of tensor muscle of 64
tympanus, tensor muscle canal of 66
tympanus, tensor muscle of 44
tympanus, umbo of 64
tympanus, vestibular (oral) window of 66
types of cells 11
types of nerve endings 56
types of neurons 56
types of skin of the human body 72
types of teeth 41

U

ulna 13, 14, 15, 16, 18, 19, 28
ulna, articular circumference of 19
ulnar vein, middle
umbilical artery 86
umbilical region 8
umbilical vein 86
uncinate process of pancreas 34, 35
unit of kidney, functional 76
urachus 74, 78
ureter 35, 74, 75, 77
ureter, female 79
ureter, male 78
ureter, opening of 77
ureter, opening of male 78
urethra 74, 77, 79, 88
urethra, corpus cavernosus of male 77
urethra, corpus spongiosum of male 78
urethra, external opening of 79
urethra, external opening of female 74, 79
urethra, external opening of male 78
urethra, female 77
urethra, internal opening of 77
urethra, internal opening of male 78
urethra, lacunae of male 77
urethra, male 77, 78
urethra, membranous part of male 78
urethra, mucous fold of male 77
urethra, navicular fossa of female 77
urethra, navicular fossa of male 77
urethra, opening of 80
urethra, opening of female 77
urethra, opening of male 77
urethra, prostatic part of male 77
urethra, prostatic sinus of prostatic part of male 77
urethra, trabeculae of corpus cavernosus of male 77
urethra, tunica albuginea of corpus cavernosus of male 77
urethra, urethral crest of prostatic area of male 77
urinary system 74–77
urinary system, female 74
urinary system, male 74
urogenital region 8
uterine gland 84
uterovaginal plexus 79
uterus 74, 79, 82, 83, 84, 85, 88
uterus during menstrual cycle 84
uterus, arrival of egg at 84
uterus, cervix of 79
uterus, entrance of oocyte in cavity of 82
uterus, fundus of 79, 82, 83, 85
uterus, implantation of morula in wall of 83
uterus, implantation of oocyte in 83
uterus, implantation of oocyte in wall of 82
uterus, isthmus of 79

uterus, ligamentum teres of 74
uterus, muscles of 89
uterus, neck (cervix) of 79, 82, 83, 85, 89
uterus, oocyte nesting in wall of 82
uterus, pathway of oocyte toward cavity of 79
uterus, phases of the genetic materia (oocyte + spermatozoon) until its implantation in the 83
uterus, stay of free egg in cavity of 84
uterus, vaginal portion of 79
uterus, vaginal portion of neck of 79
utricle of osseous labyrinth of ear 65
utricle, prostatic 74, 77
uvula 31
uvula of urinary bladder 77, 78

V

vacuole 10, 80
vagina 74, 79, 83, 85, 88
vagina, bulb of vestible of 74
vagina, opening of 74, 79
vaginal nerve 79
vagus nerve 54, 70
vallecula of epiglottis 68
valve 52
valve of a main artery 52
valve of heart, cranial view of 48
valve of Heister (gallbladder) 37
valve of illeocecal foramen 39
valve, anterior cusp of left atrioventricular 48
valve, atrioventricular 50
valve, closed 52
valve, iliocecal 31, 39
valve, inside 52
valve, left atrioventricular 48
valve, left semilunar cusp of aortic 48
valve, mitral 48, 50
valve, open 52
valve, posterior cusp of left atrioventricular 48
valve, posterior cusp of right atrioventricular 48
valve, posterior semilunar cusp of aortic 48
valve, pulmonary 50
valve, schematic section of a venous 52
valve, septal cusp of atrioventricular 48
valve, sigmoid 50
valve, tricuspid 50
vastus lateralis muscle 24, 25, 26
vastus medialis muscle 24, 26
Vater, papilla of 37
Vater-Pacini, corpuscle of 71
vein 41, 47, 52, 71
vein of circulatory system, main 47
vein of eye, anterior ciliary 59
vein of left ventricle, posterior 49
vein of penis, deep dorsal 78
vein of penis, superficial dorsal 78
vein of retina, central 59, 60
vein, adventitia of a main 52
vein, annular process of the fifth thoracic 22
vein, anterior tibial 47
vein, axillary 47
vein, azygus 49
vein, basilic 47
vein, brachiocephalic 47
vein, branch of pulmonary 46
vein, bronchial 46
vein, cephalic 47
vein, collagen fiber of a main 52
vein, common iliac 47
vein, digital 47
vein, elastic fiber of a main 52

vein, endothelium of a main 52
vein, external elastic lamina of a main 52
vein, external saphenous 47
vein, femoral 47
vein, gastroduodenal 35
vein, gastroepiploic 47
vein, great cardiac 49
vein, hepatic 47
vein, ileal 35
vein, inferior dental 41
vein, inferior mesenteric 35, 47
vein, intermediate 47
vein, internal iliac 47
vein, internal jugular 47
vein, internal saphenous 47
vein, intima of a main 42
vein, jejunal 35
vein, left brachiocephalic 49
vein, left gastric 35
vein, left pulmonary 48, 49
vein, main coronary 49
vein, medial layer of a main 52
vein, median inferior 47
vein, middle cardiac 49
vein, middle colic 35
vein, middle cubital 46
vein, ophthalmic 61
vein, ovarian 35, 75
vein, palmar 47
vein, portal 35, 37
vein, portal hepatic 47
vein, posterior tibial 47
vein, pulmonary 46, 47
vein, renal 35, 75
vein, right brachiocephalic 49
vein, right colic 35
vein, right pulmonary 48, 49
vein, schematic section of a 52
vein, schematic section of a main 52
vein, splenic 35, 40
vein, subclavian 47
vein, superior bulb of jugular 64
vein, superior mesenteric 35, 47
vein, suprarenal 75
vein, testicular 35, 75
vein, umbilical 86, 88
vein, valve of a main 52
vein, vertebral 49
vein, vorticose 60
vena cava, inferior 35, 37, 47, 49
vena cava, ligament of 37
vena cava, sinus of 49
vena cava, superior 47, 48, 49
ventricle of brain, choroid plexus of lateral 63
ventricle of myocardium, left 48
ventricle of myocardium, right 48
ventricle, choroid plexus of IV 54
ventricle, laryngeal 68
ventricle, left 49, 50
ventricle, posterior veins of posterior 49
ventricle, relaxed 50
ventricle, right 48, 49, 50
vermis of cerebellum 54
vertebra 22, 23, 88
vertebra, accesory process of the fourth lumbar 23
vertebra, annular process of the fifth thoracic 22
vertebra, apex of spinous process of the seventh cervical 23
vertebra, bifid spinous process of the fourth cervical 23
vertebra, body of 73
vertebra, costal facet of transverse process of the fifth thoracic 22
vertebra, fifth 22
vertebra, first cervical 16, 21, 25
vertebra, fourth lumbar 16, 21, 24

vertebra, Gegenbaur process of the fourth lumbar 23
vertebra, inferior articular facet of the fifth lumbar 23
vertebra, inferior articular process of the fifth lumbar 23
vertebra, inferior articular process of the sixth thoracic 22
vertebra, inferior notch of the pedicle of the fifth lumbar 23
vertebra, inferior vertebral notch of the sixth thoracic 22
vertebra, intervertebral surface of the body of the fifth thoracic 22
vertebra, intervertebral surface of the body of the fourth lumbar 23
vertebra, lamina of the fourth cervical 23
vertebra, lamina of the fourth lumbar 23
vertebra, lamina of the seventh cervical 23
vertebra, lateral view of the fifth lumbar 23
vertebra, mammillary process of the fourth lumbar 23
vertebra, mammillary tubercle of the fifth lumbar 23
vertebra, margin of the crest of the fourth lumbar 23
vertebra, pedicle of the fifth lumbar 23
vertebra, pedicle of the fifth thoracic 22
vertebra, pedicle of the fourth cervical 23
vertebra, pedicle of the fourth lumbar 23
vertebra, pedicle of the seventh cervical 23
vertebra, semilunar process of the seventh cervical 23
vertebra, seventh cervical 23
vertebra, sixth 22
vertebra, spinous process of the fifth lumbar 23
vertebra, spinous process of the fifth thoracic 22
vertebra, spinous process of the fourth lumbar 23
vertebra, spinous process of the seventh cervical 23
vertebra, spinous process of the sixth thoracic 22
vertebra, superior articular facet of the fourth cervical 23
vertebra, superior articular process of the fifth thoracic 22
vertebra, superior articular process of the fourth lumbar 23
vertebra, superior articular process of the seventh cervical 23
vertebra, superior articular process of the sixth thoracic 22
vertebra, superior articular surface of the vertebral body of the fourth cervical 23
vertebra, superior costal fossae of the sixth thoracic 22
vertebra, superior notch of pedicle of the fourth cervical 23
vertebra, superior view of the fourth cervical 23
vertebra, thoracic 11, 22, 57, 58
vertebra, transverse (costal) process of the fourth lumbar 23
vertebra, transverse articular process of the fifth lumbar 23
vertebra, transverse foramen of the fourth cervical 23

vertebra, transverse foramen of the seventh cervical 23
vertebra, transverse process of the fifth thoracic 22
vertebra, transverse process of the fourth cervical 23
vertebra, transverse process of the seventh cervical 23
vertebra, transverse process of the sixth thoracic 22
vertebra, unciform process of the fourth cervical 23
vertebra, vertebral arch of the seventh cervical 23
vertebra, vertebral body of the fifth lumbar 23
vertebra, vertebral body of the fourth cervical 23
vertebra, vertebral body of the fourth lumbar 23
vertebra, vertebral body of the seventh cervical 23
vertebra, vertebral body of the sixth thoracic 22
vertebra, vertebral foramen of the fourth cervical 23
vertebra, vertebral foramen of the fourth lumbar 23
vertebrae, cervical 14, 15, 57, 58
vertebrae, dorsal 15, 57, 58
vertebrae, lumbar 15, 57, 58
vertebrae, median fissure of 73
vertebral column of fetus, 87
vertebral column, 13, 21, 87, 88

vertebral region 9
vertical muscle of tongue 68
vesicle of a cell, secretory 10
vesicle, defferential plexus of seminal 78
vesicle, ductus deferens of seminal 78
vesicle, pinocytic 10
vesicle, seminal 74, 77, 78
vessel of epigastrus, blood 35
vessel of maternal blood 86
vessel of mucosa of wall of ileum, central lymphatic 38
vessel of retina, blood 59
vessel, blood 31
vessel, lymphatic 31
vessel, umbilical 89
vestibular nerve of osseous labyrinth of ear 65
vestibule of nose 67
vestibule of osseous labyrinth 65
vestibule of tympanic cavity, window of 66
vestibule of vagina, bulb of 74
vestibule, buccal 31, 68
vestibulocochlear nerve 54
vibration 66
view of a hepatic cell, microscopic 36
view of acinar cell of pancreas, microscopic 34
view of atlas, anterior 23
view of atlas, lateral 23
view of atlas, superior 23

view of axis, lateral 22
view of axis, posterior 22
view of brain showing visual paths, inferior 63
view of brain, inferior 54
view of brain, superior 53
view of bronchi, anterior 45
view of bronchi, posterior 45
view of bronchial pulmonary segments identified by colors, anterior 44
view of ciliary body, posterior 62
view of colon, anterior 39
view of duodenum, anterior 32, 34
view of duodenum, general 32
view of duodenum, posterior 35
view of female genitourinary system, lateral 74
view of fifth lumbar vertebra, lateral 23
view of foot, lateral 18
view of foot, posterior 18, 28
view of fourth cervical vertebra, superior 23
view of gallbladder, anterior 37
view of hand, dorsal 18, 28
view of hand, palmar 18, 28
view of head, anterior 27
view of head, lateral 27
view of heart valves, superior 48
view of heart, anterior 49
view of heart, posterior 49
view of interior wall of stomach, microscopic 32

view of iris, anterior 62
view of iris, posterior 62
view of kidney, anterior 75
view of large intestine, anterior 39
view of larynx, anterior 45
view of larynx, posterior 45
view of liver, anterior 36
view of liver, posterior 37
view of male genitourinary system, lateral 74
view of myocardium, superior 48
view of nasal bones, lateral 67
view of nasal wall, lateral 67
view of osseous labyrinth of ear, transverse 65
view of pancreas, anterior 34
view of pancreas, posterior 35
view of sacrum bone, lateral 22
view of sacrum, anterior 22
view of sacrum, lateral 22
view of skull, anterior 17
view of skull, lateral 17
view of skull, posterior 17
view of skull, superior 17
view of small intestine, anterior 38
view of spleen, anteromedial 40
view of spleen, laterocranial 40
view of spleen, schematic section of caudal 40
view of stomach, anterior 32, 33
view of stomach, general 32
view of tongue, lateral 68
view of tongue, posterior 68
view of tongue, superior 69

view of tooth, superior 41
view of trachea, anterior 44, 45
view of trachea, posterior 45
view of urinary bladder, anterior 77
view of urinary bladder, posterior 77
villi of fetus 87
villi of mucosa of upper part of small intestine, 38
villi of taste bud, taste 69
villi of wall of ileum, intestinal 38
vitamine 31
vitelline duct (yolk stalk) 87
vomer bone 12, 17, 67

W

window of cochlea 65
window of cochlea, fossa of 64, 66
window of ear, oval 66
window of tympanic cavity, vestibular 66
wing (ala) of ilium 20
wing of sacrum bone 22
Wirsung, duct of 34
Wolff, duct of 74
woman, abdomen of a pregnant 88
woman, anatomy of 7
womb of mother 86, 89
womb, maternal 86, 89
wrist 6, 7

Z

zona pellucida of ovary 81
zona pellucida, follicular cells adhering to 81

zone (area) innervated by fragments of the spinal cord, cutaneous 57
zone (area) innervated by spinal nerves, cutaneous 58
zone (area) of a spermatozoon, middle 80
zone (area) of ovary, polar 81
zone (area) of perception of acid taste, tongue 69
zone (area) of perception of salty taste, tongue 69
zone (area) of perception of sour taste, tongue 69
zone (area) of perception of sweet taste, tongue 69
zone (area) of perception of taste, tongue 69
zone (area), taste 70
zone (cortex) of kidney, cortical 75
zone of stomach, cardiac 32
zygomatic bone 12, 14, 16, 17
zygomatic region 9
zygomaticus major muscle 24, 27
zygomaticus minus muscle 24, 27
zygote 85
zygote divided into eight blastomeres 83
zygote divided into four blastomeres 83
zygote divided into two blastomeres 83
zygote, formation of 83

Contents

Synopsis 4

Foreword. 5

The Human Body. 6
1. Male Surface Anatomy 6
   Anterior view—Posterior view 6
2. Female Surface Anatomy. 7
   Anterior view—Posterior view 7
3. Regions of the Body. Anterior View 8
4. Regions of the Body. Posterior View 9

The Cell 10
1. The Human Cell. 10
2. Types of Cells. DNA 11
   *Eukaryotic cell—Prokaryotic cell—Glycogen
   —Schematic representation of the DNA molecule.* . . . 11

The Skeletal System 12
1. Bones of the Head 12
2. Bones of the Trunk and Limbs 13
3. Skeleton. Anterior View 14
4. Skeleton. Posterior View 15
5. Skeleton. Lateral View 16
6. Skull 17
   *Anterior view—Lateral view—Superior view—
   Posterior view* 17
7. Hands and Feet 18
   *Dorsal view of hand—Palmar view of hand—
   Dorsal view of foot—Lateral view of foot* 18
8. Upper Limbs. 19
   *Clavicle—Humerus—Scapula—
   Radius—Ulna* 19
9. Lower Limbs 20
   Femur—Hip bone—Tibia—Fibula—Patella. 20
10. Vertebral Column. 21
11. Vertebrae 22
    *Anterior view of sacrum—Lateral view of sacrum—
    Lateral view of axis— Posterior view of axis—Lateral
    view of sixth thoracic vertebra—Superior view of fifth
    thoracic vertebra* 22
12. Vertebrae 23
    *Anterior view of atlas—Superior view of atlas—
    Superior view of fourth cervical vertebra—
    Superior view of seventh cervical vertebra—Superior
    view of fourth lumbar vertebra—Lateral view of
    fifth lumbar vertebra* 23

The Muscular System 24
1. Muscles of the Human Body. Anterior View 24
2. Muscles of the Human Body. Posterior View. 25
3. Muscles of the Human Body. Lateral View 26
4. Muscles of the Head 27
   Anterior view—Lateral view. 27
5. Hands and Feet 28
   *Dorsal view of hand—Palmar view of hand—
   Posterior view of foot—Lateral view of foot* 28
6. Shapes and Types of Muscles and Muscular Insertions 29
   Skeletal muscle fiber of the human body 29
7. Movements of the Forearm 30

The Digestive System 31
1. Organs and Elements of the Digestive System and Steps
   of the Digestive Process 31
2. Stomach 32
   *General view of stomach and duodenum—
   Microscopic view of stomach lining showing
   gastric folds—Projection of stomach on
   anterior abdominal wall* 32
3. Stomach 33
   *Anterior view of stomach—Schematic section
   of stomach wall.* 33
4. Duodenum and Pancreas 34
   *Anterior view of duodenum and pancreas—
   Microscopic view of acinar cells of the pancreas—
   Projection of duodenum and pancreas on
   anterior abdominal wall* 34
5. Duodenum and Pancreas 35
   *Retroperitoneal organs of posterior abdominal wall
   and epigastric vessels—Posterior view of
   duodenum and pancreas* 35
6. Liver 36
   *Anterior view of liver—Microscopic view of a
   hepatic cell—Projection of liver on anterior
   abdominal wall* 36
7. Liver and Gallbladder 37
   *Posterior view of liver—Anterior view
   of gallbladder.* 37
8. Small Intestine. 38
   *Anterior view of small intestine "surrounded" by
   large intestine—Transverse section of superior part
   of small intestine—Projection of small intestine on
   anterior abdominal wall—Schematic section of
   ileum wall* 38
9. Large Intestine. 39
   *Anterior view of large intestine—Projection
   of large intestine on anterior abdominal wall
   —Frontal section of ascending colon, cecum,
   and vermiform appendix—Schematic
   section of colon wall* 39
10. Spleen 40
    *Anteromedial view—Laterosuperior view—Section
    of inferior view.* 40
11. Teeth 41
    *Deciduous teeth—Age of eruption—Longitudinal
    section of an incisor tooth—Section of mandible with
    triple root teeth—Types of teeth.* 41
12. Teeth 42
    *Permanent teeth—Age of eruption—Location of teeth
    in mouth—Schematic longitudinal section of a
    premolar tooth—Innervation of teeth.* 42

The Respiratory System. 43
1. Lungs and Bronchi 43
   Projection of bronchi on lung's wall 43
2. Location of Bronchopulmonary Segments. 44
   *Anterior view of trachea and bronchopulmonary
   segments identified by colors* 44
3. Larynx, Trachea, and Bronchi 45
   *Anterior view of larynx, trachea, and bronchi—
   Posterior view of larynx, trachea, and bronchi* 45
4. Breathing. Elements and Mechanism 46
   *Bronchiole and alveolus—Gas exchange in
   alveoli—Process of blood oxygenation developed
   in pulmonary alveolar capillaries—Mechanism of
   breathing—Inhalation—Exhalation* 46

The Circulatory System 47
1. Principal Veins and Arteries of the Circulatory
   System 47
2. Heart 48
   *Anterolateral section of heart—Superior view of
   myocardium and heart valves—Projection of heart
   on anterior abdominal wall.* 48
3. Heart 49
   Anterior view of heart—Posterior view of heart . . . 49
4. Phases of Heartbeat 50
   *Atrial diastole—Atrial systole—
   Ventricular systole—Atrial diastole.* 50
5. Blood. Blood Cells and Platelets 51
   *White blood cells—Platelet—Red blood cells—
   Microscopic view of blood cells* 51
6. Veins, Arteries, and Valves 52
   *Schematic view of a venous valve—Section
   of a main artery—Section of a main vein—
   Section of an artery—Section of a capillary—
   Section of a vein* 52

The Nervous System 53
1. Brain. Superior View. 53
2. Brain. Inferior View. 54
3. Peripheral Nervous System 55
4. Neurons. 56
   *Structure of a motor neuron—Types of neurons—
   Types of nerve endings* 56
5. Spinal Cord 57
   *Dorsal spinal cord with spinal nerves—Cutaneous
   areas innervated by fragments of the spinal cord* . . 57
6. Spinal Nerves. 58
   *Cutaneous areas innervated by spinal nerves—
   Spinal cord with spinal nerves* 58

The Senses 59
1. Sight. 59
   Sagittal section of an eye—Section of retina 59
2. Sight. 60
   Optic nerve—Back of the eye 60
3. Sight. 61
   *Lateral view of eye—Eye and eyelids—Lacrimal
   organs* 61
4. Sight. 62
   *Diagram of visual pathway—Posterior view
   of iris and ciliary body—Anterior view of iris and
   pupil* 62
5. Sight. 63
   *Inferior view of brain showing visual pathways—
   Main defects of sight and their correction* 63
1. Ear. 64
   *External acoustic meatus—External ear—Tympanic
   membrane* 64
2. Ear. 65
   *Transversal view of osseous labyrinth—Osseous
   cochlea—Osseous labyrinth* 65
3. Ear. 66
   *Development of auditory process—Ossicles of the
   middle ear—Tympanic cavity.* 66
1. Smell 67
   *Lateral view of nasal wall—Section of olfactory
   mucosa—Olfactory cell—Lateral view of nasal
   bones and cartilages—Olfactory membrane* 67
1. Taste. 68
   *Lateral view of tongue—Posterior view of tongue—
   Transverse (coronal) sections of tongue.* 68
2. Taste. 69
   *Schematic section of parts of the tongue—Superior
   view of tongue showing taste areas—Schematic*

*section of part of a circumvallate papilla—
Schematic section of a taste bud* 69
3. Taste 70
   Transmission of a taste sensation—Taste buds. . . 70
1. Touch 71
   *Schematic section of the skin—Epidermis—Tactile
   sensory receptors.* 71
2. Touch 72
   *Schematic sections of different types of skin of
   the human body—Schematic section of a hair—
   Sweat gland* 72
3. Touch 73
   *The reflex pathway for the immediate response
   of the body to painful external stimuli—
   Schematic section of a finger—The nail* 73

The Urinary System 74
1. Male Urinary System. Female Urinary System. . . . 74
   *Lateral view of male genitourinary
   system—Lateral view of the female
   genitourinary system* 74
2. Kidney 75
   *Anterior view of right kidney—Anterior
   view of left kidney—Section of anterior
   view of left kidney.* 75
3. Nephron and Glomerulus 76
   *Schematic view of a nephron, the functional
   unit of the kidney—Renal glomerulus* 76
4. Urinary Bladder and Urethra 77
   *Posterior view of urinary bladder—Anterior
   view of urinary bladder—Male urethra—
   Female urethra* 77

The Reproductive System 78
1. Male Reproductive Organs 78
   Spinal innervation—Penis 78
2. Female Reproductive Organs 79
   *Internal genitalia—External genitalia—Spinal
   innervation* 79
3. Male Reproductive Elements 80
   *External genitalia—Testis—Spermatogenesis—
   Spermatozoon* 80
4. Female Reproductive Elements 81
   *Schematic section of an ovary—Structure of
   an ovum.* 81
5. Birth and Pathway of an Unfertilized Ovum 82
   *Stages of ovum until its expulsion in
   menstruation* 82
6. Birth and Pathway of a Fertilized Ovum 83
   *Stages of the genetic material (ovum +
   spermatozoon) until its implantation in
   the uterus* 83
7. Menstrual Cycle. 84
   *Fertilization and pregnancy mechanism—
   Phases of the cycle—Ovary—Uterus—Basal
   temperature* 84
8. Fertilization 85
   *Fertilization and nesting—Fertilization of
   ovum and formation of morula* 85
9. Placenta and Development of Fetus. 86
   *Section of placenta—Development of fetus
   inside maternal womb.* 86
10. The Fetus During Gestation 87
    Elements of a four-week-old embryo 87
11. Abdomen of a Pregnant Woman 88
12. Placenta During Birth. 89
    *Placenta during the fifth week of pregnancy—
    The delivery process* 89
13. Breast and Mammary Glands. 90

Index 91